Samuel Stanhope Smith

An Essay on the Causes of the Variety of Complexion and Figure in the Human Species

Samuel Stanhope Smith

An Essay on the Causes of the Variety of Complexion and Figure in the Human Species

ISBN/EAN: 9783337780807

Printed in Europe, USA, Canada, Australia, Japan

Cover: Foto ©berggeist007 / pixelio.de

More available books at **www.hansebooks.com**

AN

ESSAY

ON THE

CAUSES OF THE VARIETY

OF

COMPLEXION AND *FIGURE*

IN THE

HUMAN SPECIES.

TO WHICH ARE ADDED

STRICTURES

ON LORD *KAIMS*'s DISCOURSE, ON THE ORIGINAL DIVERSITY OF MANKIND.

By the Reverend SAMUEL STANHOPE SMITH, *D. D.* Vice-President, and Professor of Moral Philosophy in the College of New-Jersey; and Member of the American Philosophical Society, held at Philadelphia for promoting Useful Knowledge.

PHILADELPHIA:

Printed and Sold by *ROBERT AITKEN*, at Pope's Head, Market Street.

M.DCC.LXXXVII.

THE substance of the following Essay was delivered in the annual Oration, before the Philosophical Society in Philadelphia, February 28th, 1787.—And the whole is published at the request of the Society.

AT A MEETING OF THE AMERICAN PHILOSOPHICAL SOCIETY, ON FRIDAY EVENING, THE 28th OF FEBRUARY, 1787.

On Motion Ordered,

THAT the thanks of the Society be given to the Reverend Doctor SAMUEL S. SMITH, *for his ingenious and learned Oration delivered this evening, and that he be requested to furnish the Society with a copy of the same for Publication.*

Extract from the Minutes,

JAMES HUTCHINSON,
ROBERT PATTERSON,
SAMUEL MAGAW,
JOHN FOULKE,
} *Secretaries.*

C

ESSAY

ON THE

CAUSES OF THE VARIETY

OF

COMPLEXION AND FIGURE

IN THE

HUMAN SPECIES.

IN the hiſtory and philoſophy of human nature, one of the firſt objects that ſtrikes an obſerver is the variety of complexion and of figure among mankind. To aſſign the cauſes of this phænomenon has been frequently a ſubject of curious ſpeculation. Many philoſophers have reſolved the difficulties with which this inquiry is attended by having recourſe to the arbitrary hypotheſis that men are originally ſprung from different ſtocks, and are therefore divided by nature into different ſpecies. But as we are not at liberty to make this ſuppoſition, ſo I hold it to be unphilosophical to recur to hypothesis, when the whole effect may, on proper investigation, be accounted for by the ordinary laws of nature*.

On this discussion I am now about to enter; and shall probably unfold, in its progress, some principles the full importance of which will not be obvious, at first view, to those who have not been accustomed to observe the operations of nature with minute and careful attention. Principles, however, which, experience leads me to believe, will acquire additional evidence from time and observation.

Of the causes of these varieties among mankind I shall treat under the heads—

I. Of CLIMATE.

II. Of the STATE of SOCIETY.

In treating this subject, I shall not espouse any peculiar system of medical principles which,

* It is no small objection to this hypothesis, that these species can never be ascertained. We have no means of distinguishing how many were originally formed, or where any of them are now to be found. And they must have been long since so mixed by the migrations of mankind, that the properties of each species can never be determined. Besides, this supposition unavoidably confounds the whole philosophy of human nature.—*See conclusion of this essay.*

which, in the continual revolutions of opinion, might be in hazard of being afterwards discarded. I ſhall, as much as poſſible, avoid uſing terms of art; or attempting to explain the *manner of operation* of the cauſes, where diverſity of opinion among phyſicians, has left the ſubject in doubt.

And, in the beginning, permit me to make one general remark which muſt often have occurred to every judicious inquirer into the powers both of moral and of phyſical cauſes— that every permanent and characteriſtical variety in human nature, is effected by ſlow and almoſt imperceptible gradations. Great and ſudden changes are too violent for the delicate conſtitution of man, and always tend to deſtroy the ſyſtem. But changes that become incorporated, and that form the character of a climate or a nation, are progreſſively carried on through ſeveral generations, till the cauſes that produce them have attained their utmoſt operation. In this way, the minuteſt cauſes, acting conſtantly, and long continued, will neceſſarily create great and conſpicuous differences among mankind.

I. Of the firſt claſs of cauſes, I ſhall treat under the head of climate. In

In tracing the globe from the pole to the equator, we obſerve a gradation in the complexion nearly in proportion to the latitude of the country. Immediately below the arctic circle a high and ſanguine colour prevails. From this you deſcend to the mixture of red in white. Afterwards ſucceed, the brown, the olive, the tawny, and at length the black, as you proceed to the line. The ſame diſtance from the ſun, however, does not, in every region, indicate the ſame temperature of climate. Some ſecondary cauſes muſt be taken into conſideration as correcting and limiting its influence. The elevation of the land, its vicinity to the ſea, the nature of the ſoil, the ſtate of cultivation, the courſe of winds, and many other circumſtances, enter into this view. Elevated and mountainous countries are cool in proportion to their altitude above the level of the ſea—vicinity to the ocean produces oppoſite effects in northern and ſouthern latitudes; for the ocean being of a more equal temperature than the land, in one caſe corrects the cold, in the other, moderates the heat. Ranges of mountains, ſuch as the Appenines in Italy, and Taurus, Caucaſus and Imaus in Aſia, by interrupting the courſe of cold winds, render the protected countries below

below them warmer, and the countries above them colder, than is equivalent to the proportional difference of latitude. The frigid zone in Asia is much wider than it is in Europe; and that continent hardly knows a temperate zone. From the northern ocean to Caucasus, says Montesquieu, Asia may be considered as a flat mountain. Thence to the ocean that washes Persia and India, it is a low and level country without seas, and protected by this immense range of hills from the polar winds. The Asiatic is, therefore, warmer than the European continent below the fortieth degree of latitude; and, above that latitude, is much more cold. Climate also receives some difference from the nature of the soil; and some from the degree of cultivation—Sand is susceptible of greater heat than clay; and an uncultivated region, shaded with forests, and covered with undrained marshes, is more frigid in northern, and more temperate in southern latitudes, than a country laid open to the direct and constant action of the sun. History informs that, when Germany and Scythia were buried in forests, the Romans often transported their armies across the frozen Danube; but, since the civilization of those barbarous regions, the Danube rarely freezes.

Many

Many other circumſtances might be enumerated which modify the influence of climate. Theſe will be ſufficient to give a general idea of the ſubject. And by the intelligent reader they may be eaſily extended, and applied to the ſtate of particular countries.

From the preceding obſervations we derive this concluſion, that there is a general ratio of heat and cold, which forms what we call climate, and a general reſemblance of nations, according to the latitude from the equator; ſubject however, to innumerable varieties from the infinite combinations of the circumſtances I have ſuggeſted. After having exhibited the *general* effect, I ſhall take up the capital deviations from it that are found in the world, and endeavour to ſhew that they naturally reſult from certain concurrences of theſe modifying cauſes.

Our experience verifies the power of climate on the complexion. The heat of ſummer darkens the ſkin, the cold of winter chafes it, and excites a ſanguine colour. Theſe alternate effects in the temperate zone tend in ſome degree to correct one another. But when heat or cold predominates in any region, on, it impreſſes, in the ſame proportion, a permanent and characteriſtical complexion. The degree in which it predominates may be conſidered as a conſtant cauſe to the action of which the human body is expoſed. This cauſe will affect the nerves by tention or relaxation, by dilatation or contraction—It will affect the fluids by increaſing or leſſening the perſpiration, and by altering the proportions of all the ſecretions—It will peculiarly affect the ſkin by the immediate operation of the atmoſphere, of the ſun's rays, or of the principle of cold upon its delicate texture. Every ſenſible difference in the degree of the cauſe, will create a viſible change in the human body. To ſuggeſt at preſent a ſingle example.—A cold and piercing air chafes the countenance and exalts the complexion. An air that is warm and miſty relaxes the conſtitution, and gives ſome tendency, in valetudinarians eſpecially, to a bilious hue. Theſe effects are tranſient, and interchangeable in countries where heat and cold alternately ſucceed in nearly equal proportions. But when the climate conſtantly repeats the one or the other of theſe effects in any degree, then, in proportion, an habitual colour begins to be formed. Colour and figure may

be

be ſtiled habits of the body. Like other habits, they are created, not by great and ſudden impreſſions, but by continual and almoſt imperceptible touches. Of habits both of mind and body, nations are ſuſceptible as well as individuals. They are tranſmitted to offspring, and augmented by inheritance. Long in growing to maturity, national features, like national manners become fixed, only after a ſucceſſion of ages. They become, however, fixed at laſt. And if we can aſcertain any effect produced by a given ſtate of weather or of climate, it requires only repetition during a ſufficient length of time, to augment and impreſs it with a permanent character. The ſanguine countenance will, for this reaſon, be perpetual in the higheſt latitudes of the temperate zone; and we ſhall forever find the ſwarthy, the olive, the tawny and the black, as we deſcend to the ſouth.

The uniformity of the effect in the ſame climate, and on men in a ſimilar ſtate of ſociety, proves the power and certainty of the cauſe. If the advocates of different human ſpecies ſuppoſe that the beneficent Deity hath created the inhabitants of the earth of different colours, becauſe theſe colours are beſt adapted

adapted to their reſpective zones, it ſurely places his benevolence in a more advantageous light to ſay, he has given to human nature the power of accommodating itſelf to every zone. This pliancy of nature is favourable to the unions of the moſt diſtant nations, and facilitates the acquiſition and the extenſion of ſcience which would otherwiſe be confined to few objects, and to a very limited range. It opens the way particularly to the knowledge of the globe which we inhabit; a ſubject ſo important and intereſting to man.—It is verified by experience. Mankind are forever changing their habitations by conqueſt or by commerce. And we find them in all climates not only able to endure the change, but ſo *aſſimilated* by time, that we cannot ſay with certainty whoſe anceſtor was the native of the clime, and whoſe the intruding foreigner.

I will here propoſe a few principles on the change of colour, that are not liable to diſpute, and that may tend to ſhed ſome light on this ſubject.

In the beginning, it may be proper to obſerve that the ſkin, though extremely delicate

 and

and eaſily ſuſceptible of impreſſion from external cauſes, is, from its ſtructure, among the leaſt mutable parts of the body*. Change of complexion does for this reaſon continue long, from whatever cauſe it may have ariſen. And if the cauſes of colour have deeply penetrated the texture of the ſkin, it becomes perpetual. Figures therefore, that are ſtained with paints inſerted by punctures made in its ſubſtance, can never be effaced†. An ardent ſun is able intirely to penetrate its texture. Even in our climate, the ſkin, when firſt expoſed to the direct and continued action of the ſolar rays, is inflamed into bliſters, and ſcorched through its whole ſubſtance. Such an operation not only changes its colour, but increaſes its thickneſs. The ſtimulus of heat exciting a greater flux of humours to the ſkin, tends to incraſſate its ſubſtance, till it becomes denſe enough to reſiſt the action of the exciting

* Anatomiſts inform us that, like the bones, it has few or no veſſels, and therefore is not liable to thoſe changes of augmentation or diminution, and continual alteration of parts, to which the fleſh, the blood, and whole vaſcular ſyſtem is ſubject.

† It is well known what a length of time is required to efface the freckles contracted in a fair ſkin by the expoſure of a ſingle day. Freckles are ſeen of all ſhades of colour. They are known to be created by the ſun; and become indelible by time. The ſun has power equally to change every part of the ſkin, when equally expoſed to its action. And it is, not improperly, obſerved by ſome writers that colour may be juſtly conſidered as an univerſal freckle.

citing cauſe*. On the ſame principle, friction excites bliſters in the hand of the labourer, and thickens the ſkin till it becomes able to endure the continued operation of his inſtruments. The face or the hand, expoſed uncovered during an intire ſummer, contracts a colour of the darkeſt brown. In a torrid climate, where the inhabitants are naked, the colour will be as much deeper, as the ardor of the ſun is both more conſtant and more intenſe. And if we compare the dark hue that, among us, is ſometimes formed by continual expoſure, with the colour of the African, the difference is not greater than is proportioned to the augmented heat and conſtancy of the climate†.

The principle of colour is not, however, to be derived ſolely from the action of the ſun upon the ſkin. Heat, eſpecially, when united with putrid exhalations that copiouſly impregnate the atmoſphere in warm and uncultivated regions, relaxes the nervous ſyſtem. The bile in conſequence is augmented, and ſhed

* Anatomiſts know that all people of colour have their ſkin thicker than people of a fair complexion, in proportion to the darkneſs of the hue.

† If the force of fire be ſufficient at a given diſtance, to ſcorch the fuel, approach it as much nearer as is proportional to the difference of heat between our climate and that of Africa, and it will burn it black,

ſhed through the whole maſs of the body. This liquor tinges the complexion of a yellow colour, which aſſumes by time a darker hue. In many other inſtances, we ſee that relaxation, whether it be cauſed by the vapours of ſtagnant waters, or by ſedentary occupations, or by loſs of blood, or by indolence, ſubjects men to diſorders of the bile, and diſcolours the ſkin. It has been proved, by phyſicians, that in fervid climates the bile is always augmented in proportion to the heat*. Bile expoſed to the ſun and air, is known to change its colour to black—black is therefore the tropical hue. Men who remove from northern to ſouthern regions are uſually attacked by dangerous diſorders that leave the blood impoveriſhed, and ſhed a yellow appearance over the ſkin. Theſe diſorders are perhaps the efforts of nature in breaking down and changing the conſtitution, in order to accommodate it to the climate; or to give it that degree of relaxation, and to mingle with it that proportion of bile, which is neceſſary for its new ſituation†. On this dark ground

* See Dr. M'Clurg on the bile.

† Phyſicians differ in their opinions concerning the ſtate of the bile in warm countries. Some ſuppoſe that it is thrown out to be a corrector of putridity. Others ſuppoſe that in all relaxed habits, the bile is itſelf in a putrid

ground the hue of the climate becomes, at length, deeply and permanently impreſſed.

On the ſubject of the phyſical cauſes of colour I ſhall reduce my principles to a few ſhort propoſitions derived chiefly from experience and obſervation, and placed in ſuch connexion as to illuſtrate and ſupport one another. They may be enlarged and multiplied by men of leiſure and talents who are diſpoſed to purſue the inquiry farther.

1. It is a fact that the ſun darkens the ſkin although there be no uncommon redundancy of the bile.

2. It is alſo a fact that redundancy of bile darkens the ſkin, although there be no uncommon expoſure to the ſun*.

3. It is a fact equally certain that where both

putrid ſtate. I decide not among the opinions of phyſicians. Whichever be true, the theory I advance will be equally juſt. The bile will be augmented; it will tinge the ſkin, and there, whether in a ſound or putrid ſtate, will receive the action of the ſun and atmoſphere, and be, in proportion, changed towards black.

* Redundancy of bile long continued, as in the caſe of the black jaundice, or of extreme melancholy, creates a colour almoſt perfectly black.

both cauſes co-operate, the effect is much greater, and the colour much deeper*.

4. It is diſcovered by anatomiſts that the ſkin conſiſts of three lamellæ, or folds,—the external, which in all nations is an extremely fine and tranſparent integument,—the interior, which is alſo white,—and an intermediate, which is a cellular membrane filled with a mucous ſubſtance.

5. This ſubſtance, whatever it be, is altered in its appearance and colour with every change of the conſtitution—As appears in bluſhing, in fevers, or in conſequence of exerciſe. A lax nerve, that does not propel the blood with vigour, leaves it pale and ſallow—it is inſtantly affected with the ſmalleſt ſurcharge of bile, and ſtained of a yellow colour.

6. The change of climate produces a proportionable alteration in the internal ſtate and ſtructure of the body, and in the quantity of the

* This we ſee verified in thoſe perſons who have been long ſubject to bilious diſorders, if they have been much expoſed to the ſun. Their complexion becomes in that caſe extremely dark.

the ſecretions*. In ſouthern climates particularly, the bile, as has been remarked, is always augmented.

7. Bile, expoſed to the ſun and air in a ſtagnant, or nearly in a ſtagnant ſtate, tends in its colour towards black.

8. The ſecretions as they approach the extremities, become more languid in their motion till at length they come almoſt to a fixed ſtate in the ſkin.

9. The aqueous parts eſcaping eaſily by perſpiration through the pores of the ſkin, thoſe that are more denſe and incraſſated remain in a mucous or glutinous ſtate in that cellular membrane between the interior ſkin and the ſcarf, and receive there, during a long time, the impreſſions of external and diſcolouring cauſes.

10. The bile is peculiarly liable to become mucous and incraſſated†; and in this ſtate, being

* This appears from the diſorders with which men are uſually attacked on changing their climate; and from the difference of figure and aſpect which takes place in conſequence of ſuch removals. This latter reflexion will afterwards be further illuſtrated.

† In this ſtate it is always copiouſly found, in the ſtomach and inteſtines at leaſt in conſequence of a bilious habit of body.

being unfit for perſpiration, and attaching itſelf ſtrongly to that ſpongy tiſſue of nerves, it is there detained for a length of time till it receives the repeated action of the ſun and atmoſphere.

11. From all the preceding principles taken together it appears that the complexion in any climate will be changed towards black, in proportion to the degree of heat, in the atmoſphere, and to the quantity of bile in the ſkin.

12. The vapours of ſtagnant waters with which uncultivated regions abound; all great fatigues and hardſhips; poverty and naſtineſs, tend as well as heat, to augment the bile. Hence, no leſs than from their nakedneſs, ſavages will always be diſcoloured, even in cold climates. For though cold, when aſſiſted by ſucculent nouriſhment, and by the comfortable lodging and clothing furniſhed in civilized ſociety, propels the blood with force to the extremities, and clears the complexion; yet when hardſhips and bad living relax the ſyſtem, and when poor and ſhivering ſavages, under the arctic cold, do not poſſeſs thoſe conveniencies that, by opening the pores, and

and cheriſhing the body, aſſiſt the motion of the blood to the ſurface, the florid and ſanguine principle is repelled, and the complexion is left to be formed by the dark coloured bile; which, in that ſtate, becomes the more dark, becauſe the obſtruction of the pores preſerves it longer in a fixed ſtate in the ſkin. Hence, perhaps, the deep Lapponian complexion which has been eſteemed a phænomenon ſo difficult to be explained.

13. Cold, where it is not extreme*, is followed by a contrary effect. It corrects the bile, it braces the conſtitution, it propels the blood to the ſurface of the body with vigour; and renders the complexion clear and florid†.

Such are the obſervations which I propoſe concerning the proximate cauſe of colour in the human ſpecies. But I remark, with pleaſure, that whether this theory be well founded or not, the fact may be perfectly aſcertained,

* Extreme cold is followed by an effect ſimilar to that of extreme heat. It relaxes the conſtitution by overſtraining it, and augments the bile. This, together with the fatigues and hardſhips and other evils of ſavage life, renders the complexion darker beneath the arctic circle, than it is in the middle regions of the temperate zone, even in a ſavage ſtate of ſociety.

† Cold air is known to contain a conſiderable quantity of nitre; and this ingredient is known to be favourable to a clear and ruddy complexion.

ed, that climate has all that power to change the complexion which I ſuppoſe, and which is neceſſary to the preſent ſubject.—It appears from the whole ſtate of the world—it appears from obvious and undeniable events within the memory of hiſtory, and from events even within our own view.

Encircle the earth in every zone, and, making thoſe reaſonable allowances which have been already ſuggeſted, and which will afterwards be farther explained, you will ſee every zone marked by its diſtinct and characteriſtical colour. The black prevails under the equator; under the tropics, the dark copper; and on this ſide of the tropic of Cancer, to the ſeventieth degree of north latitude, you ſucceſſively diſcern the olive, the brown, the fair and the ſanguine complexion. Of each of theſe there are ſeveral tints or ſhades. And under the arctic circle, you return again to the dark hue. This general uniformity in the effect indicates an influence in the climate that, under the ſame circumſtances, will always operate in the ſame manner. The apparent deviations from the law of climate that exiſt in different regions of the globe will be found to confirm it, when I come, in the pro-

greſs

greſs of this diſcourſe, to point out their cauſes*.

The power of climate, I have ſaid, appears from obvious and undeniable events within the memory of hiſtory. From the Baltic to the Mediterranean you trace the different latitudes by various ſhades of colour. From the ſame, or from nearly reſembling nations, are derived the fair German, the dark Frenchman, the ſwarthy Spaniard and Sicilian. The ſouth of Spain is diſtinguiſhed by complexion from the north. The ſame obſervation may be applied to moſt of the other countries of Europe. And if we would extend it beyond Europe to the great nations of the eaſt, it is applicable to Turkey, to Arabia, to Perſia and to China. The people of Pekin are fair; at Canton they are nearly black. The Perſians near the Caſpian ſea are among the faireſt people in the world†; near the gulph of Ormus they are of a dark olive. The inhabitants of the Stony and Deſert Arabia are tawny; while thoſe of Arabia the Happy are as

* Independently on the effects of the ſtate of ſociety which will be hereafter illuſtrated, there are, in reality, various climates under the ſame parallels.

† The *fair Circaſſian* has become proverbial of the women of a neighbouring nation.

as black as the Ethiopians. In these ancient nations, colour holds a regular progression with the latitude from the equator. The examples of the Chinese and the Arabians are the more decisive on this subject because they are known to have continued, from the remotest antiquity, unmingled with other nations. The latter, in particular, can be traced up to their origin from one family. But no example can carry with it greater force on this subject than that of the Jews. Descended from one stock, prohibited by their most sacred institutions from intermarrying with other nations, and yet dispersed, according to the divine predictions, into every country on the globe, this one people is marked with the colours of all. Fair in Britain and Germany, brown in France and in Turkey, swarthy in Portugal and in Spain, olive in Syria and in Chaldea, tawny or copper coloured in Arabia and in Egypt*.

Another example of the power of climate more immediately subject to our own view may be shewn in the inhabitants of these United States. Sprung within a few years from the British, the Irish and the German nations

* Buffon's nat. hist. vol. 3d.

nations who are the faireſt people in Europe, they are now ſpread over this continent from the thirty firſt to the forty fifth degree of northern latitude. And, notwithſtanding the temperature of the climate—notwithſtanding the ſhortneſs of the period ſince their firſt eſtabliſhment in America—notwithſtanding the continual mixture of Europeans with thoſe born in the country—notwithſtanding previous ideas of beauty that prompted them to guard againſt the influence of the climate—and notwithſtanding the ſtate of high civilization in which they took poſſeſſion of their new habitations, they have already ſuffered a viſible change. A certain countenance of paleneſs and of ſoftneſs ſtrikes a traveller from Britain the moment he arives upon our ſhore. A degree of ſallowneſs is viſible to him which, through familiarity, or the want of a general ſtandard of compariſon, hardly attracts our obſervation. This effect is more obvious in the middle, and ſtill more, in the ſouthern, than in the northern ſtates. It is more obſervable in the low lands near the ocean than as you approach the Apalachian mountains; and more, in the lower and labouring claſſes of people, than in families of eaſy fortune who poſſeſs the means, and

and the inclination to protect their complexion. The inhabitants of New-Jerſey, below the falls of the rivers, are ſomewhat darker in their colour than the people of Pennſylvania, both becauſe the land is lower in its ſituation, and becauſe it is covered with a greater quantity of ſtagnant water. A more ſouthern latitude augments the colour along the ſhores of Maryland and Virginia. At length the low lands of the Carolinas and of Georgia degenerate to a complexion that is but a few ſhades lighter than that of the Iroquois. I ſpeak of the poor and labouring claſſes of the people who are always firſt and moſt deeply affected by the influence of climate and who eventually give the national complexion to every country. The change of complexion which has already paſſed upon theſe people is not eaſily imagined by an inhabitant of Britain, and furniſhes the cleareſt evidence to an attentive obſerver of nature that, if they were thrown, like the native Indians, into a ſavage ſtate they would be perfectly marked, in time, with the ſame colour. Not only their complexion, but their whole conſtitution ſeems to be changed. So thin and meagre is the habit of the poor, and of the overſeers of their ſlaves, that, frequently, their

limbs

limbs appear to have a diſproportioned length to the body, and the ſhape of the ſkeleton is evidently diſcernible through the ſkin*. If theſe men had been found in a diſtant region where no memory of their origin remained, the philoſophers who eſpouſe the hypotheſis of different ſpecies of men would have produced them in proof, as they have often done nations diſtinguiſhed by ſmaller differences than diſtinguiſh theſe from their European anceſtors†. Examples taken from the natives of

* The dark colour of the natives of the Weſt-India Iſlands is well known to approach very near a dark copper. The deſcendents of the Spaniards in ſouth America are already become copper coloured: [ſee phil. tranſ. of roy. ſoc. Lond. N° 476 ſect. 4.] The Portugueſe of Mitomba in Sierra Leona on the coaſt of Africa have, by intermarrying with the natives, and by adopting their manners, become, in a few generations, perfectly aſſimilated in aſpect, figure and complexion, [ſee treatiſe on the trade of Great Britain to Africa, by an African merchant.] And lord Kaims, who cannot be ſuſpected of partiality on this ſubject, ſays of another Portugueſe ſettlement on the coaſt of Congo, that the deſcendents of thoſe poliſhed Europeans, have become, both in their perſons and their manners, more like beaſts than like men. [ſee ſketches of man, prel. diſc.] Theſe examples tend to ſtrengthen the inference drawn from the changes that have happened in the Anglo-Americans. And they ſhew how eaſily climate would aſimilate foreigners to natives in the courſe of time, if they would adopt the ſame manners, and equally expoſe themſelves to its influence.

† The habit of America is, in general, more ſlender than that of Britain. But the extremely meagre aſpect of the pooreſt and loweſt claſs of people in ſome of the ſouthern ſtates may ariſe from the following cauſe, that the changes produced by climate are, in the firſt inſtance, generally diſeaſes. Hereafter, when the conſtitution ſhall be perfectly accomodated to the climate, it will by degrees aſſume a more regular and agreeable figure. The Anglo-Americans, however, will never reſemble the native Indians. Civilization will prevent ſo great a degeneracy either in the colour

of the United States are the ſtronger becauſe climate has not had time to impreſs upon them its full character. And the change has been retarded by the arts of ſociety, and by the continual intermixture of foreign nations.

Theſe changes may, to perſons who think ſuperficially on the ſubject, ſeem more ſlow in their progreſs than is conſiſtent with the principles hitherto laid down concerning the influence of climate. But in the philoſophy of human nature it is worthy of obſervation, that all national changes, whether moral or phyſical, advance by imperceptible gradations, and are not accompliſhed but in a ſeries of ages. Ten centuries were requiſite to poliſh the manners of Europe. It is not improbable that an equal ſpace of time may be neceſſary to form the countenance, and the figure of the body—to receive all the inſenſible and infinite impreſſions of climate—to combine theſe with the effects that reſult from the ſtate of ſociety—to blend both along with perſonal peculiarities—and by the innumerable

colour or the features. Even if they were thrown back again into the ſavage ſtate the reſemblance would not be complete; becauſe, the one would receive the impreſſions of the climate on the ground of features formed in Europe—the others have received them on the ground of features formed in a very different region of the globe. The effects of ſuch various combinations can never be the ſame.

ble unions of families to melt down the whole into one uniform and national countenance*. It is even queſtiònable whether, amidſt eternal migrations and conqueſts, any nation in Europe has yet received the full effects of theſe cauſes. China and Arabia are perhaps the only civilized countries in the world in which they have attained their utmoſt operation; becauſe they are the only countries in which the people have been able, during a long ſucceſſion of ages, to preſerve themſelves unmixed with other nations. Each parallel of latitude is, among them, diſtinctly marked by its peculiar complexion. In no other nations is there ſuch a regular and perfect gradation of colour as is traced from the fair natives of Pekin, to Canton, whoſe inhabitants are of the darkeſt copper—or, from the olive of the Deſert Arabia to the deep black of the province of Yemen. It is plain then that the cauſes of colour, and of other varieties in the human ſpecies, have not yet had their full operation on the inhabitants of theſe United States. Such an operation, however, they have already had as affords a ſtrong proof,

 and

* In ſavage life men more ſpeedily receive the characteriſtic features of the climate, and of the ſtate of ſociety: becauſe the habits and ideas of ſociety among them are few and ſimple; and to the action of the climate they are expoſed naked and defenceleſs to ſuffer its full force at once.

and an interesting example of the powerful influence of climate*.

The preceding observations have been intended chiefly to explain the principle of colour. I proceed now to illustrate the influence of climate on other varieties of the human body.

It would be impossible, in the compass of a discourse like the present, to enter minutely into the description of every feature of the countenance and of every limb of the body, and to explain all the changes in each that may possibly be produced by the power of climate combined with other accidental causes. Our knowledge of the human constitution, or of the globe, or of the powers of nature is, perhaps,

* The reader will please to keep in mind that in remarking on the changes that have passed on the Anglo-Americans, I have in view the mass of the people. And that I have in view likewise natives of the second or third generation, and not such as are sprung from parents, one or both of whom have been born in Europe; though even with regard to these the remarks will be found to hold in a great degree. I am aware that particular instances may be adduced that will seem to contradict each remark. But such examples do not overthrow general conclusions derived from the body of the populace. And these instances, I am persuaded, will be very rare among those who have had a clear American descent by both parents, for two or three generations. They will be more rare in the low and level country where the climate is more different, and the descents more remote from Europe, than in the countries to the west where the land rises into hills. Here the climate is more similar to that in the middle of Europe, and the people are more mingled with emigrants from Ireland and Germany.

perhaps, not ſufficiently accurate and extenſive to enable us to offer a ſatisfactory ſolution of every difficulty that an attentive or a captious obſerver might propoſe. But if we are able, on juſt principles, to explain the capital varieties, in figure and aſpect, that exiſt among different nations, it ought to ſatisfy a reaſonable inquirer; as no minuter differences can be ſufficient to conſtitute a diſtinct ſpecies.

I ſhall, therefore, confine my obſervations at preſent, to thoſe conſpicuous varieties that appear in the hair, the figure of the head, the ſize of the limbs, and in the principal features of the face.

The hair generally follows the law of the complexion, becauſe, its roots, being planted in the ſkin, derive its nouriſhment and its colour from the ſame ſubſtance which there contributes to form the complexion. Every gradation of colour in the ſkin, from the brown to the perfectly black, is accompanied with proportionable ſhades in the hair. The pale red, or ſandy complexion, on the other hand, is uſually attended with redneſs of the hair. Between theſe two points is found almoſt

every

every other colour of this excreſcence, ariſing from the accidental mixture of the principles of black and red in different proportions. White hair, which is found only with the faireſt ſkin, ſeems to be the middle of the extremes, and the ground in which they both are blended*. The extremes, if I may ſpeak ſo, are as near to each other as to any point in the circle, and are often found to run into one another. The Highlanders of Scotland are generally either black or red. A red beard is frequently united with black hair. And if, in a red or dark coloured family, a child happens to deviate from the law of the houſe, it is commonly to the oppoſite extreme. On this obſervation permit me to remark, that thoſe who deny the identity of human origin, becauſe one nation is red and another is black, might, on the ſame principle, deny, to perſons of different complexion, the identity of family. But as the fact, in the latter inſtance, is certain; we may, in the former, reaſonably conclude that, the ſtate of nerves or fluids which contributes to produce one or other of theſe effects in a ſingle family, may be

* That black hair is ſometimes ſuppoſed to be united with the faireſt ſkin ariſes from the deception which the contraſt between the hair and ſkin puts upon the ſight.

be the general tendency of a particular climate. In this example, at leaſt, we ſee that the human conſtitution is capable of being moulded, by phyſical cauſes, into many of the varieties that diſtinguiſh mankind. It is contrary therefore to ſound philoſophy, which never aſſigns different cauſes, without neceſſity, for ſimilar events, to have recourſe, for explaining theſe varieties, to the hypotheſis of ſeveral original ſpecies*.

Climate poſſeſſes great and evident influence on the hair not only of men, but of all other animals. The changes which this excreſcence undergoes in them is at leaſt equal to what it ſuffers in man. If, in one caſe, theſe tranſmutations are acknowledged to be conſiſtent with identity of kind, they ought not, in the other, to be eſteemed criterions of diſtinct

* If we ſuppoſe different ſpecies to have been created, how ſhall we determine their number? Are any of them loſt? or where ſhall we, at preſent find them clearly diſtinguiſhed from all others? or were the ſpecies of men made capable of being blended together, contrary to the nature of other animals, ſo that they ſhould never be diſcriminated, ſo rendering the end unneceſſary for which they were ſuppoſed to be created? If we have reaſon, from the varieties that exiſt in the ſame family, or in the ſame nation, to conclude that the Danes, the French, the Turks, and people even more remote are of one ſpecies, have we not the ſame reaſon to conclude that the nations beyond them, and who do not differ from the laſt by more conſpicuous diſtinctions, than the laſt differ from the firſt, are alſo of the ſame ſpecies. By purſuing this progreſſion we ſhall find but one ſpecies from the equator to the pole.

diſtinct ſpecies. Nature hath adapted the pliancy of her work to the ſituations in which ſhe may require it to be placed. The beaver, removed to the warm latitudes, exchanges its fur, and the ſheep its wool, for a coarſe hair that preſerves the animal in a more moderate temperature. The coarſe and black ſhag of the bear is converted, in the arctic regions, into the fineſt and whiteſt fur. The horſe, the deer, and almoſt every animal protected by hair, doubles his coat in the beginning of winter, and ſheds it in the ſpring when it is no longer uſeful. The fineneſs and denſity of the hair is augmented in proportion to the latitude of the country. The Canadian and Ruſſian furs are, therefore, better than the furs of climates farther ſouth. The colour of the hair is likewiſe changed by climate. The bear is *white* under the arctic circle; and in high northern latitudes, *black* foxes are moſt frequently found. Similar effects of climate are diſcernible on mankind. Almoſt every nation is diſtinguiſhed by ſome peculiar quality of this excreſcence. The hair of the Danes is generally red, of the Engliſh fair or brown, and of the French commonly black. The Highlanders of Scotland are divided between red and

and black. Red hair is frequently found in the cold and elevated regions of the Alps, although black be the predominant complexion at the foot of thoſe mountains. The aborigines of America, like all people of colour, have black hair; and it is generally long and ſtraight. The ſtraightneſs of the hair may ariſe from the relaxation of the climate, or from the humidity of an uncultivated region. But whatever be the cauſe, the Anglo-Americans already feel its influence. And curled locks ſo frequent among their anceſtors are rare in the United States*.

Black is the moſt uſual colour of the human hair, becauſe thoſe climates that are moſt extenſive, and moſt favourable to population, tend to the dark complexion. Climates that are not naturally marked by a peculiar colour may owe the accidental predominancy of one, to the conſtitutional qualities of an anceſtral family—They may owe the prevalence of a variety of colours to the early ſettlement of different

* They are moſt rare in the ſouthern ſtates, and in thoſe families that are fartheſt deſcended from their European origin. Straight lank hair is almoſt a general characteriſtic of the Americans of the ſecond and third race. It is impoſſible, however, to predict what effect hereafter the clearing of the country and the progreſs of cultivation may have on the hair as well as other qualities of the Americans. They will neceſſarily produce a great change in the climate, and conſequently in the human conſtitution.

different families; or to the migrations or conquefts of different nations. England is, perhaps for this reafon, the country in which is feen the greateft variety in the colour of the hair.

But the form of this excrefcence which principally merits obfervation, becaufe it feems to be fartheft removed from the ordinary laws of nature, is feen in that fparfe and curled fubftance peculiar to a part of Africa, and to a few of the Afiatic iflands.

This peculiarity has been urged as a decifive character of a diftinct fpecies with more affurance than became philofophers but tolerably acquainted with the operations of nature. The fparfenefs of the African hair is analogous to the effect which a warm climate has been fhewn to have on other animals. Cold, by obftructing the perfpiration tends to throw out the perfpirable matter accumulated at the fkin in an additional coat of hair. A warm climate, by opening the pores, evaporates this matter before it can be concreted into the fubftance of hair; and the laxnefs and aperture of the pores renders the hair liable

able to be eaſily eradicated by innumerable accidents.

Its curl may reſult in part, perhaps, from external heat, and in part from the nature of the ſubſtance or ſecretion by which it is nouriſhed. That it depends in a degree on the quality of the ſecretion is rendered probable from its appearance on the chin, and on other parts of the human body. Climate is as much diſtinguiſhed by the nature and proportion of the ſecretions as by the degree of heat. Whatever be the nutriment of the hair it ſeems to be combined, in the torrid zone of Africa, with ſome fluid of a highly volatile or ardent quality. That it is combined with a ſtrong volatile ſalt, the rank and offenſive ſmell of many African nations, gives us reaſon to ſuſpect. Saline ſecretions tend to curl and to burn the hair. The evaporation of any volatile ſpirit would render its ſurface dry and diſpoſed to contract, while the center continuing diſtended by the vital motion, theſe oppoſite dilatations and contractions would neceſſarily produce a curve, and make the hair grow involved. This conjecture receives ſome confirmation by obſerving that the negroes born in the United States of America are gradually loſing the ſtrong ſmell of the African zone; their hair is, at the ſame time, growing leſs involved, and becoming denſer and longer*.

External and violent heat parching the extremities of the hair tends likewiſe to involve it. A hair held near the fire inſtantly coils itſelf up. The herbs roll up their leaves, in the extreme heats of ſummer, during the day, and expand them again in the coolneſs of the evening. Africa is the hotteſt country on the globe. The ancients who frequented the Aſiatic zone eſteemed the African an uninhabitable zone of fire. The hair as well as the whole human conſtitution ſuffers, in this region, the effects of an intenſe heat.

The manners of the people add to the influence of the climate. Being ſavages they have few arts to protect them from its intenſity. The heat and ſerenity of the ſky preſerving the life of children without much care of

* Many negroes of the third race in America have thick cloſe hair, extended to four or five inches in length. In ſome who take great pains to comb and dreſs it in oil, it is even longer, and they are able to extend it into a ſhort queue. This is particularly the caſe with ſome domeſtic ſervants who have more leiſure and better means than others to cheriſh their hair. Many negroes, however, cut their hair as faſt as it grows, preferring it ſhort.

of the parent, they ſeem to be the moſt negligent people of their offspring in the univerſe*. Able themſelves to endure the extremes of that ardent climate, they inure their children from their moſt tender age. They ſuffer them to lie in the aſhes of their huts, or to roll in the duſt and ſand beneath the direct rays of a burning ſun. The mother, if ſhe is engaged, lays down the infant on the firſt ſpot ſhe finds, and is ſeldom at the pains to ſeek the miſerable ſhelter of a barren ſhrub, which is all that the interior country affords. Thus the hair is criſped, while the complexion is blackened by exceſſive heat†. There is

* The manners of a people are formed, in a great meaſure, by their neceſſities. The dangers of the North-American climate render the natives uncommonly attentive to the preſervation of their children. The African climate not laying its ſavage inhabitants under any neceſſity to be careful, they expoſe their children to its utmoſt influence without concern.

† I have myſelf been witneſs of this treatment of children by the ſlaves in the ſouthern ſtates where they are numerous enough to retain many of their African cuſtoms. I ſpeak of the field ſlaves who, living in little villages on their plantations at a diſtance from their maſters' manſions, are ſlow in adopting the manners of their ſuperiors. There I have ſeen the mother of a child, within leſs than ſix weeks after it was born, take it with her to the field and lay it in the ſand beneath a hot ſun while ſhe hoed her corn-row down and up. She would then ſuckle it a few minutes and return to her work, leaving the child in the ſame expoſure, although ſhe might have gained, within a few yards, a convenient ſhade. Struck at firſt with the apparent barbarity of this treatment I have remonſtrated with them on the ſubject; and was uniformly told that dry ſand and a hot ſun were never found to hurt them. This treatment tends to add to the injury that the climate does to the hair. A ſimilar negligence among the poor, who ſuffer their children to lie in aſhes, or on the naked ground, and

is probably a concurrence of both the preceding causes in the production of the effect. The influence of heat either external, or internal, or of both, in giving the form to the hair of the Africans, appears, not only from its sparseness and its curl, but, from its colour. It is not of a shining, but an adust black, and its extremities tend to brown as if it had been scorched by the fire.

Having treated so largely on the form of this excrescence in that country, where it deviates farthest from the common law of the species. I proceed to consider a few of the remaining varieties among mankind.

The whole of the Tartar race are of low stature—Their heads have a disproportioned magnitude to the rest of the body—Their shoulders are raised, and their necks are short—Their eyes are small, and appear by the jutting of the eyebrows over them, to be sunk in the head—The nose is short, and rises but little from the face—The check is elevated and spread out on the sides—The whole features

and who expose them without covering for their heads to the sun and wind, we find greatly injures their hair. We rarely see persons who have been bred in extreme poverty, who have it not short, and thin, and frittered. But the heat of the sand and of the sun in Africa must have a much more powerful effect.

tures are remarkably coarſe and deformed: And all theſe peculiarities are aggravated, as you proceed towards the pole, in the Lapponian, Borandian and Samoiede races, which, as Buffon juſtly remarks, are Tartars reduced to the laſt degree of degeneracy.—A race of men reſembling the Laplanders we find in a ſimilar climate in America. The frozen countries round Hudſon's bay are, except Siberia, the coldeſt in the world. And here the inhabitants are between four and five feet in height—Their heads are large—Their eyes are little and weak—And their hands, feet, and whole limbs uncommonly ſmall.

Theſe effects naturally reſult from extreme cold. Cold contracts the nerves, as it does all ſolid bodies. The inhabitants grow under the conſtriction of continual froſt as under the forcible compreſſion of ſome powerful machine. Men will therefore be found in the higheſt latitudes, forever ſmall and of low ſtature*. The exceſſive rigors of theſe frozen regions affect chiefly the extremities. The blood circulating to them with a more languid

* A moderate degree of cold is neceſſary to give force and tone to the nerves, and to raiſe the human body to its largeſt ſize. But extreme cold overſtrains and contracts them. Therefore theſe northern tribes are not only ſmall, but weak and timid.

languid and feeble motion has not ſufficient vigour to reſiſt the impreſſions of the cold. Theſe limbs conſequently ſuffer a greater contraction and diminution than the reſt of the body. But the blood flowing with warmth and force to the breaſt and head, and perhaps with the more force, that its courſe to the extremities is obſtructed, diſtends theſe parts to a diſproportionate ſize. There is a regular gradation in the effect of the climate, and in the figure of the people from the Tartars to the tribes round Hudſon's bay. The Tartars are taller and thicker than the Laplanders or the Samoiedes, becauſe their climate is leſs ſevere—The northern Americans are the moſt diminutive of all, their extremities are the ſmalleſt, and their breaſt and head of the moſt diſproportioned magnitude, becauſe, inhabiting a climate equally ſevere with the Samoiedes, they are reduced to a more ſavage ſtate of ſociety*.

Extreme

* The neighbourhood of the Ruſſians, of the Chineſe, and even of the Tartars who have adopted many improvements from the civilized nations that border upon them, give the Laplanders and Siberians conſiderable advantages over the northern Americans who are in the moſt abject ſtate of ſavage life, and totally deſtitute of every art either for convenience or protection. The principles ſtated above apply to all theſe nations in proportion to the degree of cold combined with the degree of ſavageneſs. The inhabitants of the northern civilized countries of Europe are generally of lower ſtature than thoſe in the middle regions. But civilization and a milder climate prevent them from degenerating equally with the northern Aſiatics and Americans.

Extreme cold likewiſe tends to form the next peculiarities of theſe races, their high ſhoulders, and their ſhort necks. Severe froſt prompts men to raiſe their ſhoulders as if to protect the neck, and to cheriſh the warmth of the blood that flows to the head. And the habits of an eternal winter will fix them in that poſition.—The neck will appear ſhortened beyond its due proportion, not only becauſe it ſuffers an equal contraction with the other parts of the body; but becauſe the head and breaſt being increaſed to a diſproportioned ſize, will encroach upon its length; and the natural elevation of the ſhoulders will bury what remains ſo deep as to give the head an appearance of reſting upon them for its ſupport. That theſe peculiarities are the effect of climate*, the examples produced by French miſſionaries in China, of moſt reſpectable characters, leave us no room to doubt, who aſſure us that they have ſeen, even in the forty eighth degree of northern latitude,

* As climate is often known peculiarly to affect certain parts of the body, philoſophy, if it were neceſſary, could find no more difficulty in accounting for the ſhort necks of the Tartars, and other northern tribes, as a diſeaſe of the climate, than ſhe finds in giving the ſame account for the thick necks ſo frequently found in the regions of the Alps. But the obſervations before made will probably convince the attentive reader that there is no need to reſort to ſuch a ſolution of the phænomenon, when it ſeems ſo eaſily to be explained by the known operation of natural cauſes.

latitude, the poſterity of Chineſe families who had become perfect Tartars in their ſigure and aſpect; and that they were diſtinguiſhed, in particular, by the ſame ſhortneſs of the neck, and by the ſame elevation of the ſhoulders*.

That coarſe and deformed features are the neceſſary production of the climate cannot have eſcaped the attention of the moſt incurious obſerver.—Let us attend to the effects of extreme cold. It contracts the aperture of the eyes—it draws down the brows—it raiſes the cheek—by the preſſure of the under jaw againſt the upper it diminiſhes the face in length and ſpreads it out at the ſides—and diſtorts the ſhape of every feature.

This, which is only a tranſient impreſſion in our climate, ſoon effaced by the conveniencies of ſociety, and by the changes of the ſeaſon, becomes a heightened and permanent effect in thoſe extreme regions, ariſing from the greater intenſity, and the conſtant action of the cauſe. The naked and defenceleſs condition of the people augments its violence—and beginning its operation from infancy when the features are moſt tender and ſuſceptible

* See Recueil 24 des lettres edifiantes.

tible of impreſſion, and continuing it, without remiſſion, till they have attained their utmoſt growth, they become fixed at length in the point of greateſt deformity, and form the character of the Hudſon or Siberian countenance.

The principal peculiarities that may require a farther illuſtration are the ſmallneſs of the noſe, and depreſſion of the middle of the face —the prominence of the forehead—and the extreme weakneſs of the eyes.

The middle of the face is that part which is moſt expoſed to the cold, and conſequently ſuffers moſt from its power of contraction. It firſt meets the wind, and it is fartheſt removed from the ſeat of warmth in the head. But a circumſtance of equal, or, perhaps, of greater importance on this ſubject, is that the inhabitants of frozen climates naturally drawing their breath more through the noſe, than through the mouth*, thereby direct the greateſt impulſe of the air on that feature, and the parts adjacent. Such a continual ſtream of

 air

* A froſty air inhaled by the mouth chills the body more than when it is received by the noſtrils; probably becauſe a greater quantity enters at a time. Nature therefore prompts men to keep the mouth cloſed during the prevalence of intenſe froſt.

air augments the cold, and by increaſing the contraction of the parts, reſtrains the freedom of their growth*.

Hence, likewiſe, will ariſe an eaſy ſolution of the next peculiarity, the prominence of the forehead. The ſuperior warmth and force of life in the brain that fills the upper part of the head, will naturally increaſe its ſize, and make it overhang the contracted parts below.

Laſtly the eyes in theſe rigorous climates are ſingularly affected. By the projection of the eye-brows, they appear to be ſunk into the head; the cold naturally diminiſhes their aperture; and the intenſity of the froſt concurring with the glare of eternal ſnows, ſo overſtrains theſe tender organs, that they are always weak, and the inhabitants are often liable to blindneſs at an early age.

In the temperate zone on the other hand, and in a point rather below than above the middle region of temperature, the agreeable warmth

* On the ſame principle the mercury in a thermometer may be contracted and ſunk into the bulb, by directing upon it a conſtant ſtream of air from a pair of bellows, if the bulb be frequently touched during the operation with any fluid that by a ſpeedy evaporation tends to increaſe the cold.

warmth of the air diſpoſing the nerves to the moſt free and eaſy expanſion, will open the features and increaſe the orb of the eye*. Here a large full eye, being the tendency of nature, will grow to be eſteemed a perfection. And in the ſtrain of Homer, βοωπὶς πότνια Ἥρη would convey to a Greek an idea of divine beauty that is hardly intelligible to an inhabitant of the north of Europe. All the principles of the human conſtitution unfolding themſelves freely in ſuch a region, and nature acting without conſtraint will be there ſeen moſt nearly in that perfection which was the original deſign and idea of the Creator†.

II. Having endeavoured to aſcertain the power of climate in producing many varieties in the human ſpecies, I proceed to illuſtrate the influence of the ſtate of ſociety.

On

* It is perhaps worthy of remark, that, in the three continents, the temperate climates, and eternal cold border ſo nearly upon one another that we paſs almoſt inſtantly from the former to the latter. And we find the Laplander, the Samoiede, the Mongou, and the tribes round Hudſon's bay in the neighbourhood of the Swede, the Ruſſian, the Chineſe, and the Canadian. Without attention to this remark haſty reaſoners will make the ſudden change of features in theſe nations an objection againſt the preceding philoſophy.

† It may perhaps gratify my countrymen to reflect that the United States occupy thoſe latitudes that have ever been moſt favourable to the beauty of the human form. When time ſhall have accommodated the conſtitution to its new ſtate, and cultivation ſhall have meliorated the climate, the beauties of Greece and Circaſſia may be renewed in America; as there are not a few already who rival thoſe of any other quarter of the globe.

On this ſubject I obſerve,

1. In the firſt place, that the effect of climate is augmented by a ſavage ſtate, and corrected by a ſtate of civilization. And

2. In the next place, that by the ſtate of ſociety many varieties in the human perſon are intirely formed.

In the firſt place, the effect of climate is augmented by a ſavage ſtate of ſociety and corrected by a ſtate of civilization.

A naked ſavage, ſeldom enjoying the protection of a miſerable hut, and compelled to lodge on the bare ground and under the open ſky, imbibes the influence of the ſun and atmoſphere at every pore. He inhabits an uncultivated region filled with ſtagnant waters, and covered with putrid vegetables that fall down and corrupt on the ſpot where they have grown. He pitches his wigwam on the ſide of a river, that he may enjoy the convenience of fiſhing as well as of hunting. The vapour of rivers, the exhalations of marſhes, and the noxious effluvia of decaying vegetables, fill the whole atmoſphere in an unimproved country,

country, and tend to give a dark and bilious hue to the complexion*. And the ſun acting immediately on the ſkin in this ſtate will neceſſarily impreſs a deep colour.

This effect is augmented by the practice of painting, to which ſavages are often obliged to have recourſe in order to protect themſelves from the impreſſions of the humid earth on which they lie, or of a noxious atmoſphere to which they are expoſed without covering. Painting taken up at firſt through neceſſity is afterwards employed as an ornament; and a ſavage is ſeldom ſeen without having his ſkin covered with ſome compoſition that ſpoils the fineneſs of its texture, and impairs the beauty and clearneſs of its natural colour. This is known to be the effect of the fineſt paints and waſhes

* The foreſts in uncultivated countries abſorb a great part of theſe putrid vapours, otherwiſe they would be contagious and mortal. But as nature never makes her work perfect, but leaves the completion of her ſchemes to exerciſe the induſtry and wiſdom of man, the growing vegetables do not abſorb the whole effluvia of the decaying, and of the noxious marſhes that overſpread the face of ſuch a region. Nothing but civilization and culture can perfectly purify the atmoſphere. Uncultivated as well as warm countries therefore naturally tend to a bilious habit, and a dark complexion. It may ſeem an objection againſt this obſervation, that in America we often find bilious diſorders augmented in conſequence of cutting down the timber, and extending the plantations. The reaſon of which probably is that the indolence or neceſſities of a new country frequently lead men to clear the ground without draining the marſhes; or ſmall plantations are ſurrounded by unimproved foreſts. Thus, the vegetables that abſorbed the noxious moiſture being removed, it is left to fall in greater abundance on man.

waſhes that are uſed for the ſame purpoſe in poliſhed ſociety. Much more will it be the effect of thoſe coarſe and filthy unguents which are employed by ſavages. And as we ſee that coloured marks impreſſed by punctures in the ſkin become indelible, it is reaſonable to believe that the particles of paints inſinuated into its texture by forcible and frequent rubbing will tend, in like manner, to create a dark and permanent colour.

To this may be added that the frequent ſumigations by which they are obliged to guard againſt the annoyance of innumerable inſects in undrained and uncultivated countries; and the ſmoke with which their huts unſkilfully built, and without chimneys, are eternally filled, contribute to augment the natural darkneſs of the ſavage complexion. Smoke we perceive diſcolours the ſkin of thoſe labourers and mechanics who are habitually immerſed in it—it ſtains every object long expoſed to its action, by entering the pores, and adhering ſtrongly to the ſurface.—It inſinuates itſelf in a ſimilar manner into the pores of the ſkin, and there tends to change the complexion, on the ſame principles that it is changed by inſerted paints.

And

And laſtly, the hardſhips of their condition that weaken and exhauſt the principle of life—their ſcanty and meagre fare which wants the ſucculence and nouriſhment which give freſhneſs and vigour to the conſtitution—the uncertainty of their proviſion which ſometimes leaves them to languiſh with want, and ſometimes enables them to overſtrain themſelves by a ſurfeit—and their intire inattention to perſonal and domeſtic cleanlineſs, all have a prodigious effect to darken the complexion, to relax and emaciate the conſtitution, and to render the features coarſe and deformed. Of the influence of theſe cauſes we have an example in perſons reduced to extreme poverty, who are uſually as much diſtinguiſhed by their thin habit, their uncouth features, and their ſwarthy and ſqualid aſpect as by the meanneſs of their garb. Nakedneſs, expoſure, negligence of appearance, want of cleanlineſs, bad lodging, and meagre diet, ſo diſcolour and injure their form as to enable us to frame ſome judgment of the degree in which ſuch cauſes will contribute to augment the influence of climate in ſavage life. Independently on climate, theſe cauſes will render it impoſſible that a ſavage ſhould ever be fair. And the co-operation of both, will uſually render men

in

in that ſtate of ſociety extremely dark in their complexion. And generally they will be more coarſe and hard in their features and leſs robuſt in their perſons, than men who enjoy with temperance the advantages of civilized ſociety*.

As

* One of the greateſt difficulties with which a writer on this ſubject has to combat, is the ignorance and ſuperficial obſervation of the bulk of travellers who travel without the true *ſpirit of remark*. The firſt objects that meet their view in a new country and among a new people, ſeize their fancy and are recited with exaggeration; and they ſeldom have judgment and impartiality ſufficient to examine and reaſon with juſtneſs and caution; and from innumerable facts which neceſſarily have many points of difference among themſelves, to draw general concluſions. Such concluſions, when moſt juſtly drawn, they think they have refuted when they diſcover a ſingle example that ſeems not to coincide with them. In reaſonings of this kind there are few perſons who ſufficiently conſider that, however accurately we may inveſtigate cauſes and effects, our limited knowledge will always leave particular examples that will ſeem to be exceptions from any general principle.—To apply theſe remarks.—A few examples perhaps may occur, among ſavages, of regular and agreeable features, or of ſtrong and muſcular bodies; as in civilized ſociety we meet with ſome rare inſtances of aſtoniſhing beauty. If, by chance, a perſon of narrow obſervation, and incomprehenſive mind, have ſeen two or three examples of this kind, he will be ready on this ſlender foundation, to contradict the general remark, I have made concerning the coarſe and uncouth features of ſavages, and their want of thoſe fine and muſcular proportions, if I may call them ſo, in the human body, that indicate ſtrength combined with ſwiftneſs. Yet, it is certain that the general countenance of ſavage life is much more uncouth and coarſe, more unmeaning and wild, as will afterwards be ſeen when I come to point out the cauſes of it than the countenance of poliſhed ſociety: And the perſon is more ſlender, and rather fitted for the chace, than robuſt and capable of force and labour.—An American Indian, in particular, is commonly ſwift; he is rarely very ſtrong. And it has been remarked, in the many expeditions which the people of theſe ſtates have undertaken againſt the ſavages, that, in cloſe quarters, the ſtrength of an Anglo-American is uſually ſuperior to that of an Indian of the ſame ſize. The muſcles, likewiſe, on which the fine proportions of perſon ſo much depend, are generally ſmaller and more lax, than they are in improved ſociety that is not corrupted by luxury, or debilitated by ſedentary

occupations

As a ſavage ſtate contributes to augment the influence of climate; or, at leaſt, to exhibit its worſt effects upon the human conſtitution; a ſtate of civilization, on the other hand, tends to correct it, by furniſhing innumerable means of guarding againſt its power. The conveniencies of clothing and of lodging—the plenty, and healthful quality of food—a country drained, cultivated, and freed from noxious effluvia—improved ideas of beauty—the conſtant ſtudy of elegance, and the infinite arts for attaining it, even in perſonal figure and appearance, give cultivated an immenſe advantage over ſavage ſociety in its attempts to counteract the influence of climate, and to beautify the human form.

2. I come now to obſerve, what is of much more importance on this part of the ſubject, that

occupations—Their limbs, therefore, though ſtraight, are leſs beautifully turned.—A deception often paſſes on the ſenſes in judging of the beauty of ſavages—and deſcription is often more exaggerated than the ſenſes are deceived. We do not expect beauty in ſavage life. When, therefore, we happen to perceive it, the contraſt with the uſual condition of that ſtate impoſes on the mind. And the exalted repreſentations of ſavage beauty, which we ſometimes read, are true only by compariſon with ſavages.—There is a difference, in this reſpect between man, and many of the inferior animals which were intended to run wild in the foreſt. They are always the moſt beautiful when they enjoy their native liberty and range. They decay and droop when attempted to be domeſticated or confined. But man, being deſigned for ſociety and civilization, attains, in that ſtate, the greateſt perfection of his form, as well as of his whole nature.

that all the features of the human countenance are *modified*, and its intire *expreſſion* radically formed, by the ſtate of ſociety.

Every object that impreſſes the ſenſes, and every emotion that riſes in the mind, affects the features of the face the index of our feelings, and contributes to form the infinitely various countenance of man. Paucity of ideas creates a vacant and unmeaning aſpect. Agreeable and cultivated ſcenes compoſe the features, and render them regular and gay. Wild, and deformed, and ſolitary foreſts tend to impreſs on the countenance, an image of their own rudeneſs. Great varieties are created by diet and modes of living. The delicacies of refined life give a ſoft and elegant form to the features. Hard fare, and conſtant expoſure to the injuries of the weather, render them coarſe and uncouth. The infinite attentions of poliſhed ſociety give variety and expreſſion to the face. The want of intereſting emotions leaving its muſcles lax and unexerted, they are ſuffered to diſtend themſelves to a larger and groſſer ſize, and acquire a ſoft unvarying ſwell that is not diſtinctly marked by any idea. A general ſtandard of beauty has its effect in forming the human countenance

nance and figure. Every paſſion, and mode of thinking has its peculiar expreſſion—And all the preceding characters have again many variations according to their degrees of ſtrength, according to their combinations with other principles, and according to the peculiarities of conſtitution or of climate that form the ground on which the different impreſſions are received. As the degrees of civilization, as the ideas, paſſions, and objects of ſociety in different countries, and under different forms of government are infinitely various, they open a boundleſs field for variety in the human countenance. It is impoſſible to enumerate them.—They are not the ſame in any two ages of the world.—It would be unneceſſary to enumerate them, as my object is not to become a phyſiognomiſt, but to evince the poſſibility of ſo many differences exiſting in one ſpecies; and to ſuggeſt a proper mode of reaſoning on new varieties as they may occur to our obſervation.

For this purpoſe, I ſhall, in the firſt place, endeavour, by ſeveral facts and illuſtrations to evince, that the ſtate of ſociety has a great effect in varying the figure and complexion of mankind.

I ſhall

I ſhall then ſhew in what manner ſome of the moſt diſtinguiſhing features of the ſavage, and particularly of the American ſavage with whom we are beſt acquainted, naturally reſult from the rude condition in which they exiſt.

To evince that the ſtate of ſociety has a great effect in varying the figure and complexion of mankind, I ſhall derive my firſt illuſtration from the ſeveral claſſes of men in poliſhed nations. And then I ſhall ſhew that men in different ſtates of ſociety have changed, and that they have it continually in their power to change, in a great degree, the aſpect of the ſpecies, according to any general ideas or ſtandard of human beauty which they may have adopted.

1. And in the firſt place, between the ſeveral claſſes of men in poliſhed nations, who may be conſidered as people in different ſtates of ſociety, we diſcern great and obvious diſtinctions, ariſing from their ſocial habits, ideas and employments.

The poor and labouring part of the community are uſually more ſwarthy and ſqualid in their complexion, more hard in their features,

tures, and more coarſe and ill-formed in their limbs, than perſons of better fortune, and more liberal means of ſubſiſtence. They want the delicate tints of colour, the pleaſing regularity of feature, and the elegance and fine proportions of perſon. There may be particular exceptions. Luxury may disfigure the one—a fortunate coincidence of circumſtances may give a happy aſſemblage of features to the other. But theſe exceptions do not invalidate the general obſervation*. Such diſtinctions become more conſiderable by time, after families have held for ages the ſame ſtations in ſociety. They are moſt conſpicuous in thoſe countries in which the laws have made the moſt complete and permanent diviſion of ranks. What an immenſe difference exiſts, in Scotland, between the chiefs and the commonalty of the highland clans? If they had been ſeparately found in different countries, the philoſophy of ſome writers would have ranged them in different ſpecies. A ſimilar diſtinction takes place between the nobility and

* It ought to be kept in mind through the whole of the following illuſtrations that, when mention is made of the ſuperior beauty and proportions of perſons in the higher claſſes of ſociety, the remark is general. It is not intended to deny that there exiſt exceptions both of deformity among the great, and of beauty among the poor. And thoſe only are intended to be deſcribed who enjoy their fortune with temperance; becauſe luxury and exceſs tend equally with extreme poverty, to debilitate and disfigure the human conſtitution.

and peaſantry of France, of Spain, of Italy, of Germany. It is even more conſpicuous in many of the eaſtern nations, where a wider diſtance exiſts between the higheſt and the loweſt claſſes in ſociety. The *naires* or nobles of Calicut, in the Eaſt-Indies, have, with the uſual ignorance and precipitancy of travellers, been pronounced a different race from the populace; becauſe the former elevated by their rank, and devoted only to martial ſtudies and atchievments, are diſtinguiſhed by that manly beauty and elevated ſtature ſo frequently found with the profeſſion of arms, eſpecially when united with nobility of deſcent; the latter, poor and laborious, and expoſed to hardſhips, and left, by their rank, without the ſpirit or the hope to better their condition, are much more deformed and diminutive in their perſons; and in their complexion, much more black. In France, ſays Buffon, you may diſtinguiſh by their aſpect not only the nobility from the peaſantry, but the ſuperior orders of nobility from the inferior, theſe from citizens, and citizens from peaſants. You may even diſtinguiſh the peaſants of one part of the country from thoſe of another according to the fertility of the ſoil, or the nature of its product. The ſame obſervation

vation has been made on the inhabitants of different counties in England. And I have been aſſured by a moſt judicious and careful obſerver that the difference between the people in the eaſtern, and thoſe in the weſtern countries in Scotland, is ſenſible and ſtriking. The farmers who cultivate the fertile countries of the Lothians have a fairer complexion, and a better figure, than thoſe who live in the weſt, and obtain a more coarſe and ſcanty ſubſiſtence from a barren ſoil*.

If,

* It is well known that coarſe and meagre food is ever accompanied in mankind with hard features and a dark complexion. Every change of diet, and every variety in the manner of preparing it has ſome effect on the human conſtitution. A ſervant now lives in my family who was bound to me at ten years of age. Her parents were in abject poverty. The child was, in conſequence, extremely ſallow in her complexion, ſhe was emaciated, and as is common to children who have lain in the aſhes and dirt of miſerable huts, her hair was frittered and worn away to the length of little more than two inches. This girl has by a fortunate change in her mode of living, and indeed by living more like my own children than like a ſervant, become, in the ſpace of four years, freſh and ruddy in her complexion, her hair is long and flowing, and ſhe is not badly made in her perſon. A ſimilar inſtance is now in the family of a worthy clergyman, a friend and neighbour of mine. And many ſuch inſtances of the influence of diet, and modes of living will occur to a careful and attentive obſerver. It equally affects the inferior animals. The horſe, according to his treatment, may be infinitely varied in ſhape and ſize. The fleſh of many ſpecies of game differs both in taſte and colour according to the nature of the grounds on which they have fed. The fleſh of hares that have fed on high lands is much fairer than of thoſe that have fed in vallies and on damp grounds. And evdry keeper of cattle knows how much the firmneſs and flavour of the meat depends upon the manner of feeding. Nor is this unaccountable. For as each element has a different effect on the animal ſyſtem; and as the elements are combined in various proportions in different kinds of food, the means of ſubſiſtence will neceſſarily have a great influence on the human figure and complexion.——The difference, however,

If, in England, there exiſts leſs difference between the figure and appearance of perſons in the higher and lower claſſes of ſociety, than is ſeen in many other countries of Europe, it is becauſe a more general diffuſion of liberty and wealth has reduced the different ranks more nearly to a level. Science and military talents open the way to eminence and to nobility. Encouragements to induſtry, and ideas of liberty, favour the acquiſition of fortune by the loweſt orders of citizens—And, theſe not being prohibited, by the laws or cuſtoms of the nation from aſpiring to connections with the higheſt ranks, families in that country are frequently blended. You often find in citizens the beautiful figure and complexion of the nobleſt blood; and, in noble houſes, the coarſe features that were formed in lower life.

Such diſtinctions are, as yet, leſs obvious in America, becauſe, the people enjoy a greater equality; and the frequency of migration has not permitted any ſoil, or ſtate of local manners,

however, between the common people in the eaſtern and weſtern countries of Scotland, in ſeveral counties in England, and in other nations, ariſes, perhaps, not only from their food, and the ſoil which they inhabit, but, in part likewiſe, from their occupations, as huſbandmen, mechanics, or manufacturers. Huſbandry has generally a happier effect on perſonal appearance, than the ſedentary employments of manufacture.

ners, to impreſs its character deeply on the conſtitution. Equality of rank and fortune, in the citizens of the United States, ſimilarity of occupations, and of ſociety, have produced ſuch uniformity of character, that, hitherto, they are not ſtrongly marked by ſuch differences of feature as ariſe ſolely from ſocial diſtinctions. And yet there are beginning to be formed, independently on climate, certain combinations of features, the reſult of ſocial ideas, that already ſerve, in a degree, to diſtinguiſh the ſtates from one another. Hereafter they will advance into more conſiderable and characteriſtic diſtinctions.

If the white inhabitants of America afford us leſs conſpicuous inſtances, than ſome other nations, of the power of ſociety, and of the difference of ranks, in varying the human form, the blacks, in the ſouthern republics, afford one that is highly worthy the attention of philoſophers.—It has often occurred to my own obſervation.

The field ſlaves are badly fed, clothed and lodged. They live in ſmall huts on the plantations where they labour, remote from the ſociety and example of their ſuperiors.

Living by themſelves, they retain many of the cuſtoms and manners of their African anceſtors. The domeſtic ſervants, on the other hand, who are kept near the perſons, or employed in the families of their maſters, are treated with great lenity, their ſervice is light, they are fed, and clothed like their ſuperiors, they ſee their manners, adopt their habits, and inſenſibly receive the ſame ideas of elegance and beauty. The field ſlaves are, in conſequence, ſlow in changing the aſpect and figure of Africa. The domeſtic ſervants have advanced far before them in acquiring the agreeable and regular features, and the expreſſive countenance of civilized ſociety.— The former are frequently ill ſhaped. They preſerve, in a great degree, the African lips, and noſe, and hair. Their genius is dull, and their countenance ſleepy and ſtupid— The latter are ſtraight and well proportioned; their hair extended to three, four, and, ſometimes even, to ſix or eight inches; the ſize and ſhape of the mouth handſome, their features regular, their capacity good, and their look animated*. Another

* The features of the negroes in America have nndergone a greater change than the complexion; becauſe depending more on the ſtate of ſociety than on the climate, they are ſooner ſuſceptible of alteration, from its emotions, habits and ideas. This is ſtrikingly verified in the field and domeſtic

Another example of the power of ſociety is well known to every man acquainted with the ſavage tribes diſperſed along the frontiers of theſe republics. There you frequently ſee perſons who have been captivated from the ſtates, and grown up, from infancy to middle age, in the habits of ſavage life. In that time, they univerſally contract ſuch a ſtrong reſemblance of the natives in their countenance, and even in their complexion, as to afford a ſtriking proof that the differences which exiſt, in the ſame latitude, between the Anglo-American and the Indian, depend principally on the ſtate of ſociety*.

The

domeſtic ſlaves. The former, even in the third generation, retain, in a great degree, the countenance of Africa. The noſe though leſs flat, and the lips though leſs thick than in the native Africans, yet are much more flat and thick than in the family ſervants of the ſame race. Theſe have the noſe raiſed, the mouth and lips of a moderate ſize, the eyes lively and ſparkling, and often the whole compoſition of the features extremely agreeable. The hair grows ſenſibly longer in each ſucceeding race; eſpecially in thoſe who dreſs and cultivate it with care. After many inquiries, I have found that, wherever the hair is ſhort and cloſely curled in negroes of the ſecond or third race, it is becauſe they frequently cut it, to ſave themſelves the trouble of dreſſing. The great difference between the domeſtic and field ſlaves, gives reaſon to believe that, if they were perfectly free, enjoyed property, and were admitted to a liberal participation of the ſociety, rank and privileges of their maſters, they would change their African peculiarities much faſter.

* The reſemblance between theſe captives, and the native ſavages is ſo ſtrong, as at firſt to ſtrike every obſerver with aſtoniſhment. Being taken in infancy, before ſociety could have made any impreſſions upon them, and ſpending in the ſolitude and rudeneſs of ſavage life that tender and forming age, they grow up with the ſame apathy of countenance, the

ſame

The college of New-Jerſey furniſhes, at preſent, a counterpart to this example. A young Indian, now about fifteen years of age, was brought from his nation a number of years ago to receive an education in this inſtitution. And from an accurate obſervation of him during the greater part of that time, I have received the moſt perfect conviction that the

ſame

ſame lugubrious wildneſs, the ſame ſwelling of the features and muſcles of the face, the ſame form and attitude of the limbs, and the ſame characteriſtic gait, which is a great elevation of the feet when they walk, and the toe ſomewhat turned in, after the manner of a duck. Growing up perfectly naked, and expoſed to the conſtant action of the ſun and weather, amidſt all the hardſhips of the ſavage ſtate, their colour becomes very deep. As it is but a few ſhades lighter than that of the natives, it is, at a ſmall diſtance, hardly diſtinguiſhable. This example affords another proof of the greater eaſe with which a dark colour can be impreſſed, than effaced from, a ſkin originally fair. The cauſes of colour are *active* in their operation, and ſpeedily make a deep impreſſion. White is the ground on which this operation is received. And a white ſkin is to be preſerved only by protecting it from the action of theſe cauſes. Protection has merely a *negative* influence, and muſt therefore be ſlow in its effects; eſpecially as long as the ſmalleſt degree of *poſitive* agency is ſuffered from the original cauſes of colour. And as the ſkin retains with great conſtancy, impreſſions once received, all dark colours will, on both accounts, be much leſs mutable than the fair complexion. That period of time, therefore, which would be ſufficient in a ſavage ſtate, to change a white ſkin to the darkeſt hue the climate can impreſs, would, with the moſt careful protection, lighten a black colour, only a few ſhades. And becauſe this poſitive and active influence produces its effect ſo much more ſpeedily and powerfully than the negative influence that conſiſts merely in guarding againſt its operation; and ſince we ſee that the ſkin retains impreſſions ſo long, and the tanning incurred by expoſing it one day to the ſun, is not, in many days, to be effaced, we may juſtly conclude that a dark colour once contracted, if it be expoſed but a few days in the year to the action of the ſun and weather, will be many ages before it can be intirely effaced. And unleſs the difference of climate be ſo conſiderable as to operate very great changes on the internal conſtitution and to alter the whole ſtate of the ſecretions, the negroe colour, for example, may, by the expoſure of a poor and ſervile ſtate, be rendered almoſt perpetual.

ſame ſtate of ſociety, united with the ſame climate, would make the Anglo-American and the Indian countenance very nearly approximate. He was too far advanced in ſavage habits to render the obſervation complete, becauſe, all impreſſions received in the tender and pliant ſtate of the human conſtitution before the age of ſeven years, are more deep and permanent, than in any future, and equal period of life. There is an obvious difference between him and his fellow-ſtudents in the largeneſs of the mouth, and thickneſs of the lips, in the elevation of the cheek, in the darkneſs of the complexion, and the contour of the face. But theſe differences are ſenſibly diminiſhing. They ſeem, the faſter, to diminiſh in proportion as he loſes that vacancy of eye, and that lugubrious wildneſs of countenance peculiar to the ſavage ſtate, and acquires the agreeable *expreſſion* of civil life. The expreſſion of the eye, and the ſoftening of the features to civilized emotions and ideas, ſeems to have removed more than half the difference between him and us. His colour, though it is much lighter than the complexion of the native ſavage, as is evident from the ſtain of bluſhing, that, on a near inſpection, is inſtantly diſcernible, ſtill forms the principal

principal diſtinction*. There is leſs difference between his features and thoſe of his fellow-ſtudents, than we often ſee between perſons in civilized ſociety. After a careful attention to each particular feature, and compariſon of it with the correſpondent feature in us, I am now able to diſcover but little difference. And yet there is an obvious difference in the whole countenance. This circumſtance has led me to conclude that the varieties among mankind are much leſs than they appear to be. Each ſingle trait or limb, when examined apart, has, perhaps, no diverſity that may not be eaſily accounted for from known and obvious cauſes. Particular differences are ſmall. It is the reſult of the whole that ſurprizes us, by its magnitude. The combined effect of many minute varieties, like the product ariſing from the multiplication of many ſmall numbers, appears great and unaccountable. And we have not patience, or ſkill it may be, to divide this combined reſult into its leaſt portions, and to ſee, in that ſtate, how eaſy it is of comprehenſion or ſolution.

The ſtate of ſociety comprehends diet, clothing, lodging, manners, habits, face of the country,

* See the preceding note for a reaſon why the complexion is leſs changed than many of the features.

country, objects of ſcience, religion, intereſts, paſſions and ideas of all kinds, infinite in number and variety. If each of theſe cauſes be admitted to make, as undoubtedly they do, a ſmall variation on the human countenance, the different combinations and reſults of the whole muſt neceſſarily be very great; and combined with the effects of climate will be adequate to account for all the varieties we find among mankind*.

Another origin of the varieties ſpringing from the ſtate of ſociety is found in the power which men poſſeſs over themſelves of producing great changes in the human form, according to any common ſtandard of beauty which they may have adopted. The ſtandard of human beauty, in any country, is a general idea formed from the combined effect of climate and of the ſtate of ſociety. And it

* As all theſe principles may be made to operate in very different ways, the effect of one may, often, be counteracted, in a degree, by that of another. And climate will eſſentially change the effects of all. The people in different parts of the ſame country may, from various combinations of theſe cauſes, be very different. And, from the variety of combination, the poor of one country may have better complexion, features and proportions of perſon, than thoſe in another, who enjoy the moſt favourable advantages of fortune. Without attention to theſe circumſtances, a haſty obſerver will be apt to pronounce the remarks in the eſſay to be ill-founded, if he examines the human form in any country by the effect that is ſaid to ariſe from one principle alone, and do not, at the ſame time, take in the concomitant or correcting influence of other cauſes.

it reciprocally contributes to increaſe the effect from which it ſprings. Every nation varies as much from others in ideas of beauty as in perſonal appearance. Whatever be that ſtandard, there is a general effort to attain it, with more or leſs ardor and ſucceſs, in proportion to the advantages which men poſſeſs in ſociety, and to the eſtimation in which beauty is held.

To this object tend the infinite pains to compoſe the features, and to form the attitudes of children, to give them the gay and agreeable countenance that is created in company, and to exſtinguiſh all deforming emotions of the paſſions. To this object tend many of the arts of poliſhed life. How many drugs are ſold, and how many applications are made for the improvement of beauty? how many artiſts of different kinds live upon this idea of beauty? If we dance, beauty is the object; if we uſe the ſword, it is more for beauty than defence. If this general effort after appearance ſometimes leads the decrepid and deformed into abſurdity, it has, however, a great and national effect.— Of its effect in creating diſtinctions among nations in which different ideas prevail and different means are employed for attaining them, we may frame ſome conception, from the diſtinctions that exiſt in the ſame nation, in which ſimilar ideas and ſimilar means are uſed, only in different degrees. What a difference is there between the ſoft and elegant tints of complexion in genteel life, and the coarſe ruddineſs of the vulgar?—between the uncouth features and unpliant limbs of an unpoliſhed ruſtic, and the complacency of countenance, the graceful and eaſy air and figure of an improved citizen?—between the ſhaped and meaning face of a well bred lady, and the ſoft and plump ſimplicity of a country girl?—We now eaſily account for theſe differences, becauſe they are familiar to us, or, becauſe we ſee the operation of the cauſes. But if we ſhould find an intire nation diſtinguiſhed by one of theſe characters, and another by the contrary, ſome writers would pronounce them different races; although a true philoſopher ought to underſtand that the cultivation of oppoſite ideas of beauty muſt have a greater effect in diverſifying the human countenance, than various degrees, or modes, of cultivating the ſame ideas. The countenance of Europe was more various, three centuries ago, than it is at preſent. The diverſities that depend upon this cauſe are inſenſibly wearing away as the progreſs of refinement is gradually approximating the manners and ideas of the people to one ſtandard. But the influence of a general idea, or ſtandard, of the human form; and the pains taken, or the means employed, to bring our own perſons to it, are through their familiarity often little obſerved. The means employed by other nations, who aim at a different idea, attract more notice by their novelty.—The nations beyond the Indus, as well as the Tartars, from whom they ſeem to have derived their ideas of beauty with their origin*, univerſally admire ſmall eyes and large ears. They are at great pains, therefore, to compreſs their eyes at the corners, and to ſtretch their ears by heavy weights appended to them, by drawing them frequently with the hand, and by cutting their rims, ſo that they may hang down to their ſhoulders, which they conſider as the higheſt beauty. On the ſame principle, they extirpate the hair from their bodies;

* It is probable that the countries of India and China might have been peopled before the regions of Tartary; but, the frequent conqueſts which they have ſuffered, and particularly the former, from Tartarian nations, have changed their habits, ideas and perſons, even more perhaps than Europe was changed by the deluge of barbarians that overwhelmed it in the fifth century. The preſent nations beyond the Indus are, in effect, Tartars changed by the power of climate, and of a new ſtate of ſociety.

bodies; and, on the face, they leave only a few tufts here and there which they ſhave. The Tartars often extirpate the whole hair of the head, except a knot on the crown, which they braid and adorn in different manners. Similar ideas of beauty with regard to the eyes, the ears and the hair; and ſimilar cuſtoms, in the Aborigines of America, are no inconſiderable proofs that this continent has been peopled from the north-eaſtern regions of Aſia*. In Arabia and Greece large eyes are eſteemed beautiful; and in theſe countries they take extraordinary pains to ſtretch the lids, and extend their aperture. In India, they dilate the forehead in infancy, by the application of broad plates of lead. In China they compreſs the feet. In Caffraria, and many other parts of Africa, and in Lapland, they

* The celebrated Dr. Robertſon, in his hiſtory of America, deceived by the miſinformation of haſty or ignorant obſervers, has ventured to aſſert that the natives of America have no hair on their face or on their body; and like many other philoſophers, has ſet himſelf to account for a fact that never exiſted. It may be laid down almoſt as a general maxim, that the firſt relations of travellers are falſe. They judge of appearances in a new country under the prejudices of ideas and habits contracted in their own. They judge from particular inſtances, that may happen to have occurred to them, of the ſtature, the figure and the features of a whole nation. Philoſophers ought never to admit a fact on the relations of travellers, till their characters for intelligence and accurate obſervation be well aſcertained; nor even then, till the obſervation has been repeated, extended, and compared in many different lights, with other facts. The Indians have hair on the face and body; but from a falſe ſenſe of beauty they extirpate it with great pains. And traders among them are well informed, that tweezers for that purpoſe, are profitable articles of commerce.

they flatten the nofe in order to accomplifh a capricious idea of beauty. The fkin, in many nations is darkened by art; and all favages efteem certain kinds of deformity to be perfections; and ftrive to heighten the admiration of their perfons, by augmenting the wildnefs of their features. Through every country on the globe we might proceed in this manner, pointing out the many arts which the inhabitants practife to reach fome favourite idea of the human form. Arts that infenfibly, through a courfe of time, produce a great and confpicuous effect. Arts which are ufually fuppofed to have only a perfonal influence; but which really have an operation on pofterity alfo. The procefs of nature in this is as little known as in all her other works. The effect is frequently feen. Every remarkable change of feature that has grown into a habit of the body, is tranfmitted with other perfonal properties, to offspring. The coarfe features of labouring people, created by hardfhips, and by long expofure to the weather, are communicated.—The broad feet of the ruftic, that have been fpread by often treading the naked ground; and the large hand and arm, formed by conftant labour, are difcernible in children. The increafe or diminution of of any other limb or feature formed by habits that aim at an idea of beauty, may, in like manner, be imparted. We continually ſee the effect of this principle on the inferior animals. The figure, the colour and properties of the horſe are eaſily changed according to the reigning taſte. Out of the ſame original ſtock the Germans who are ſettled in Pennſylvania, raiſe large and heavy horſes; the Iriſh raiſe ſuch as are much lighter and ſmaller. According to the pains beſtowed, you may raiſe from the ſame race, horſes for the ſaddle and horſes for the draught. Even the colour can be ſpeedily changed according as faſhion is pleaſed to vary its caprice. And, if taſte preſcribes it, the fineſt horſes ſhall, in a ſhort time, be black, or white, or bay*. Human nature much more pliant, and affected by a greater variety of cauſes from food, from clothing, from lodging and from manners, is ſtill more eaſily ſuſceptible of change, according to any general ſtandard, or idea of the human form. To this principle, as well as to the manner of living, it may be, in part, attributed that the Germans, the Swedes and the French, in different parts of the United States, who live chiefly among themſelves, and

* By chuſing horſes of the requiſite qualities, to ſupply the ſtuds.

and cultivate the habits and ideas of the countries from which they emigrated, retain, even in our climate, a ſtrong reſemblance of their primitive ſtocks. Thoſe, on the other hand, who have not confined themſelves to the contracted circle of their countrymen, but have mingled freely with the Anglo-Americans, entered into their manners, and adopted their ideas, have aſſumed ſuch a likeneſs to them, that it is not eaſy now to diſtinguiſh from one another people who have ſprung from ſuch different origins.

I have ſaid that the proceſs of nature in this, as in all her other works, is inexplicable. One ſecondary cauſe, however, may be pointed out, which, ſeems to have conſiderable influence on the event*. Connexions in marriage will generally be formed on this idea of human beauty in any country. An influence

* Beſides this, men will ſoon diſcover thoſe kinds of diet, and thoſe modes of living that will be moſt favourable to their ideas. The power of imagination in pregnant women, might perhaps deſerve ſome conſideration on this ſubject. Some years ſince, this principle was carried to exceſs. I am ready to believe that philoſophers, at preſent, run to extremes on the other hand. They deny intirely the influence of imagination. But ſince the emotions of ſociety have ſo great an influence, as it is evident they have, in forming the countenance; and ſince the reſemblance of parents is communicated to children, why ſhould it be deemed incredible that thoſe general ideas which contribute to form the features of the parent, ſhould contribute alſo to form the features of the child.

influence this which will gradually approximate the countenance towards one common ſtandard. If men in the affair of marriage, were as much under management as ſome other animals, an abſolute ruler might accompliſh, in his dominions almoſt any idea of the human form. But, left as this connexion is to the paſſions and intereſts of individuals, it is more irregular and imperfect in its operations. And the negligence of the vulgar, ariſing from their want of taſte, impedes, in ſome degree, the general effect. There is however a common idea which men inſenſibly to themſelves, and almoſt without deſign, purſue. And they purſue it with more or leſs ſucceſs in proportion to the rank and taſte of different claſſes in ſociety, where they do not happen in particular inſtances, to be governed in connexions of marriage by intereſt ever void of taſte. The ſuperior ranks will always be firſt, and, in general, moſt improved, according to the prevalent idea of national beauty; becauſe, they have, it more than others, in their power to form matrimonial connexions favourable to this end. The Perſian nobility, improved in their idea of beauty, by their removal to a new climate, and a new ſtate of ſociety, have, within a few races, almoſt

almoſt effaced the characters of their Tartarian origin. The Tartars, from whom they are deſcended, are among the moſt deformed and ſtupid nations upon earth. The Perſians by obtaining the moſt beautiful and agreeable women from every country, are become a tall, and well featured, and ingenious nation. The preſent nations of Europe have with the reſinement of their manners and ideas changed and refined their perſons. Nothing can exceed the pictures of barbariſm and deformity given us of their anceſtors, by the Roman writers. Nothing can exceed the beauty of many of the preſent women of Europe and America who are deſcended from them. And the Europeans, and Americans are, the moſt beautiful people in the world, chiefly, becauſe their ſtate of ſociety is the moſt improved. Such examples tend to ſhew how much the varieties of nations may depend on ideas created by climate, adopted by inheritance, or formed by the infinite changes of ſociety and manners*. They ſhew, likewiſe how

* Society in America is gradually advancing in refinement: and if my obſervation has been juſt the preſent race furniſhes more women of exquiſite beauty than the laſt, though they may not always be found in the ſame families. And if ſociety ſhould continue its progreſſive improvement, the next race may furniſh more than the preſent. Europe has certainly made great advances in refinement of ſociety, and probably in beauty.

how much the human race might be improved both in perſonal and in mental qualities, by a well-directed care.

The ancient Greeks ſeem to have been the people moſt ſenſible of its influence. Their cuſtoms, their exerciſes, their laws, and their philoſophy, appear to have had in view, among other objects, the beauty and vigour of the human conſtitution. And it is not an improbable conjecture, that the fine models exhibited, in that country, to ſtatuaries and painters, were one cauſe of the high perfection to which the arts of ſculpture and painting arrived in Greece. If ſuch great improvements were introduced by art into the human figure, among this elegant and ingenious people, it is a proof at once of the influence of general ideas, and of how much might be effected by purſuing a juſt ſyſtem upon this ſubject. Hitherto, it has been abandoned too much to the government of chance. The great and noble have uſually had it more in their power than others to ſelect the beauty of nations in marriage: and thus, while, L without

ty. And if exact pictures could have been preſerved of the human countenance and form in every age ſince the great revolution made by the barbarians, we ſhould, perhaps, find Europe as much improved in its features as in its manners.

without ſyſtem or deſign, they gratified only their own taſte, they have generally diſtinguiſhed their order, as much by elegant proportions of perſon, and beautiful features, as by its prerogatives in ſociety. And the tales of romances that deſcribe the ſuperlative beauty of captive princeſſes, and the fictions of poets, who characteriſe their kings and nobles, by uncommon dignity of carriage and elegance of perſon, and by an elevated turn of thinking, are not to be aſcribed ſolely to the venality of writers prone to flatter the great, but have a real foundation in nature*. The ordinary ſtrain of language, which is borrowed from nature, vindicates this criticiſm. A *princely* perſon, and a *noble* thought, are uſual figures of ſpeech†.—Mental capacity, which is as various as climate, and as perſonal

* Coincident with the preceding remarks on the nations of Europe, is an obſervation made by Capt. Cook, in his laſt voyage, on the iſland Ohwyhee, and on the iſlands in general, which he viſited in the great ſouth ſea. He ſays, "the ſame ſuperiority which is obſervable in the "*Erees* [or nobles] through all the other iſlands, is found alſo here. Thoſe "whom we ſaw, were, without exception perfectly well formed; where-"as the lower ſort, beſides their general inferiority, are ſubject to all the "variety of make and figure that is ſeen in the *populace* of other countries." Cook's third voyage, book 3d, chap. 6th.

† Such is the deference paid to beauty, and the idea of ſuperiority it inſpires, that to this quality, perhaps, does the body of princes and nobles, collectively taken, in any country, owe great part of their influence over the populace. Riches and magnificence in dreſs and equipage, produce much

perſonal appearance, is, equally with the latter, ſuſceptible of improvement, from ſimilar cauſes. The body and mind have ſuch mutual influence, that whatever contributes to change the human conſtitution in its form or aſpect, has an equal influence on its powers of reaſon and genius. And theſe have again a reciprocal effect in forming the countenance. One nation may, in conſequence of conſtitutional peculiarities, created more, perhaps, by the ſtate of ſociety, than by the climate, be addicted to a grave and thoughtful philoſophy; another may poſſeſs a brilliant and creative imagination; one may be endowed with acuteneſs and wit; another may be diſtinguiſhed for being phlegmatic and dull. *Bœotian* and *Attic* wit was not a fanciful, but real diſtinction, though the remote origin of Cadmus and of Cecrops was the ſame. The ſtate of manners and ſociety in thoſe republics produced this difference more than the Bœotion air, to which it has been ſo often attributed. By the alteration of a few political, or civil, or commercial inſtitutions, and conſequently,

much of their effect by giving an artificial beauty to the perſon. How often does hiſtory remark that young princes have attached their ſubjects, and generals their ſoldiers, by extraordinary beauty? And young and beautiful queens have ever been followed and ſerved with uncommon enthuſiaſm.

ſequently, of the objects of ſociety and the train of life, the eſtabliſhment of which depended on a thouſand accidental cauſes, Thebes might have become Athens, and Athens Thebes. Different periods of ſociety, different manners, and different objects, unfold and cultivate different powers of the mind. Poetry, eloquence, and philoſophy ſeldom flouriſh together in their higheſt luſtre. They are brought to perfection by various combinations of circumſtances, and are found to ſucceed one another in the ſame nation at various periods, not becauſe the race of men, but becauſe manners and objects are changed. If as faithful a picture could be left to poſterity of perſonal as of mental qualities, we ſhould probably find the one, in theſe ſeveral periods, as various as the other; and we ſhould derive from them a new proof of the power of ſociety to multiply the varieties of the human ſpecies. Not only deficiency of objects to give ſcope to the exerciſe of the human intellect is unfavourable to its improvement; but all rudeneſs of manners is unfriendly to the culture, and the exiſtence of taſte, and even coarſe and meagre food may have ſome tendency to blunt the powers of genius. Theſe cauſes have a more powerful operation

operation than has hitherto been attributed to them by philoſophers; and merit a more minute and extenſive illuſtration than the ſubject of this diſcourſe will admit. The mental capacities of ſavages, for theſe cauſes, are uſually weaker than the capacities of men in civilized ſociety*. The powers of their minds, through defect of objects to employ them, lie dormant, and even become extinct. The faculties which, on ſome occaſions, they are found to poſſeſs, grow feeble through want of motives to call forth their exerciſe. The coarſeneſs of their food, and the filthineſs of their manners tend to blunt their genius. And the Hottentots, the Laplanders, and the people of New-Holland are the moſt ſtupid of mankind for this, among other reaſons, that they approach, in theſe reſpects, the neareſt to the brute creation†.

I am

* The exaggerated repreſentations which we ſometimes receive of the ingenuity and profound wiſdom of ſavages, are the fruits of weak and ignorant ſurprize. And ſavages are praiſed by ſome writers for the ſame reaſon that a monkey is—a certain imitation of the actions of men in ſociety, which was not expected from the rudeneſs of their condition. There are doubtleſs degrees of genius among ſavages as well as among civilized nations; but the compariſon ſhould be made of ſavages among themſelves; and not of the genius of a ſavage, with that of a poliſhed people.

† It is well known that the Africans who have been brought to America, are daily becoming, under all the diſadvantages of ſervitude, more ingenious and ſuſceptible of inſtruction. This effect, which has been taken notice of more than once, may, in part perhaps, be attributed to a change in their modes of living, as well as to ſociety, or climate.

I am now come to ſhew in what manner the features of ſavage life are affected by the ſtate of ſociety.

Civilization creates ſome affinity in countenance among all poliſhed nations. But there is ſomething ſo peculiar and ſo ſtupid in the general countenance of ſavages, that they are liable to be conſidered as an inferior grade in the deſcent from the human to the brute creation. As the civilized nations inhabit chiefly the temperate climates, and ſavages, except in America, the extremes of heat and cold, theſe differences in point of climate, combined with thoſe that neceſſarily ariſe out of their ſtate of ſociety, have produced varieties ſo great as to aſtoniſh haſty obſervers, and haſty philoſophers.—The varieties indeed produced in the features by ſavage life are great; but the real ſum of them is not ſo great as the apparent. For the eye taking in at one view, not only the actual change made in each feature, but their multiplied and mutual relations to one another, and to the whole; and each new relation giving the ſame feature a different aſpect, by compariſon, the final reſult appears prodigious*.—For example, a change made in the eye,

* See pages 63 and 64.

eye, produces a change in the whole countenance; becauſe it preſents to us, not ſingly the difference that has happened in that feature, but all the differences that ariſe from its combinations with every feature in the face. In like manner, a change in the complexion preſents not its own difference only, but a much greater effect by a ſimilar combination with the whole countenance. If both the eyes and the complexion be changed in the ſame perſon, each change affecting the whole features, the combination of the two reſults will produce a third incomparably greater than either. If, in the ſame way, we proceed to the lips, the noſe, the cheeks, and to every ſingle feature in the viſage, each produces a multiplied effect, by compariſon with the whole, and the reſult of all, like the product of a geometrical ſeries, is ſo much beyond our firſt expectation, that it confounds common obſervers, and will ſometimes embarraſs the moſt diſcerning philoſophers, till they learn, in this manner, to divide and combine effects.

To treat this ſubject fully, it would be neceſſary, in the firſt place, to aſcertain the general countenance of ſavage ſociety—and then,

then, as there are degrees in the ſavage as well as in the civilized ſtate, to diſtinguiſh the ſeveral modifications which each degree makes in the general aſpect—and, in the laſt place, to conſider the almoſt boundleſs varieties that ariſe from combining theſe general features with the effects of climate and of other cauſes already mentioned.—I do not propoſe, however, to purſue the ſubject to ſuch extent. I ſhall endeavour only to draw the general outlines of the ſavage countenance as it is formed by the ſtate of ſociety; and ſhall leave its changes reſulting from the different degrees of that ſtate, and from the combinations of theſe with other cauſes and effects, to exerciſe the leiſure and obſervation of the ingenious.

The eye of a ſavage is vacant and unexpreſſive—The whole compoſition of his countenance, is fixed and ſtupid—and over theſe unmeaning features is thrown an air of wildneſs and melancholy—The muſcles of the face are ſoft and lax—and the face is dilated at the ſides—the mouth is large—the lips ſwelled and protruded—and the noſe, in the ſame proportion, depreſſed*.

This

* In this repreſentation of the ſavage countenance, I have chiefly in view the American ſavage; although its general lineaments, and the cauſes aſſigned for them, may, in a great degree, be univerſally applied.

This is the picture.—To explain it I obſerve, that the expreſſion of the eye, and of the whole countenance depends on the nature and variety of thought and emotion. Joy and grief, ſolitude and company, objects of attention, habits, manners, whatever occupies the mind, tends to impreſs upon the countenance its peculiar traits. Mechanical occupations, and civil profeſſions, are often diſtinguiſhed by peculiarities in manner and aſpect. We frequently diſcriminate with eaſe religious denominations by a certain countenance formed by the habits of their profeſſion. Every thought has an influence in forming and diverſifying the character of the countenance, and vacuity of thought leaves it unmeaning and fixed. The infinite variety of ideas and emotions in civilized ſociety, will give every claſs of citizens ſome diſtinguiſhing expreſſion, according to their habits and occupations; and will beſtow on each individual ſome ſingular and perſonal traits, according to his genius, education, or purſuits. Between ſavage and civilized ſociety there will be all the difference that can ariſe from thinking and from want of thought. Savages will have all that uniformity among themſelves in the ſame climate, that ariſe from vacancy of mind, and want of emotion. Knowledge is various, but ignorance is ever the ſame. A vacant eye, a fixed and unmeaning countenance of idiotiſm, ſeem to reduce the ſavage in his aſpect many grades nearer than the citizen, to the brute creation. The ſolitude in which he lives, diſpoſes him to melancholy. He ſeldom ſpeaks or laughs. Society rarely enlivens his features. When not engaged in the chace, having no object to rouſe him, he reclines ſluggiſhly on the ground, he wanders careleſsly through the foreſt, or he ſits for hours in one poſture, with his eyes fixed to a ſingle point, and his ſenſes loſt in ſullen and unmeaning reverie. Theſe ſolitary and melancholy emotions ſerve to caſt over his viſage, which other cauſes render fixed, and unexpreſſive, a ſad and lugubrious air. The wild ſcenes of nature in an uncultivated country impreſs ſome reſemblance of themſelves on the features—and the paſſions of war and rage, which are almoſt the only ones that occupy the mind of a ſavage, mingle with the whole an aſpect of brutal ferocity*.

Paucity

* The inhabitants of the numerous ſmall iſlands in the great Southern and Pacific oceans form an exception to this remark. Prevented, by their iſolated ſtate, from engaging in perpetual hoſtilities with neighbouring tribes, like the continental ſavages, they are diſtinguiſhed by an air of mildneſs and complacence which is never ſeen upon the continent.

Paucity of ideas, ſolitude and melancholy, contribute likewiſe in no ſmall degree, to form the remaining features of a ſavage—a large and protruded mouth, a dilated face, and a general laxneſs and ſwell of all its muſcles*.

Society and thought put a ſtricture upon the muſcles of the face, which, while it gives them meaning and expreſſion, prevents them from dilating and ſwelling as much as they would naturally do. They collect the countenance more towards the center, and give it a greater elevation there†. But the vacant mind of the ſavage leaving the face, the index of ſentiment and paſſion, unexerted, its muſcles are relaxed, they conſequently ſpread at the ſides, and render the middle of the face broad.

Grief, peculiarly affects the figure of the lips, and makes them ſwell.—So do all ſolitary

* That theſe are natural tendencies of ſolitude, and vacancy of thought, we may diſcern by a ſmall attention to ourſelves, during a ſimilar ſtate, or ſimilar emotions of mind.

† The advancement of ſociety and knowledge is probably one reaſon why the Europeans in general have a more elevated countenance than the Aſiatics. The reader will be kind enough to remember that all remarks of this nature are only general, and not intended to reach every particular inſtance, or to inſinuate that there may not, in the infinite variety of nature, be many particular exceptions.

litary and melancholy emotions. When, therefore, theſe are the natural reſult of the ſtate of ſociety—when they operate from infancy, and are ſeldom counteracted by the more gay and intenſe emotions of civil life, the effect will at length become conſiderable. The mouth of a ſavage will generally be large, and the lips, in a leſs or greater degree thick and protruded*.

The noſe affects, and is affected by the other features of the face. The whole features uſually bear ſuch relation to one another, that if one be remarkably enlarged, it is accompanied with a proportional diminution of others. A prominent noſe is commonly connected with a thin face, and thin lips. On the other hand, a broad face, thick lips, or a large and a blunt chin, is accompanied with a certain depreſſion of the feature of the noſe. It ſeems as if the extenſion of the nerves in one direction, reſtrained and ſhortened them in another†. Savages, therefore,

* The ruſtic ſtate, by its ſolitude and want of thought and emotion, bears ſome analogy to the ſavage. And we ſee it accompanied by ſimilar effects on the viſage. The countenance vacant, the lips thick, the face broad and ſpread, and all its muſcles lax and ſwelling.

† By a ſmall experiment on ourſelves we may render this effect obvious. By a protuſion of the lips, or by drawing down the mouth at the corners, we

fore, commonly have this feature more ſunk and flat, than it is ſeen in civil ſociety. This, though a partial, is not the whole' cauſe of that extreme flatneſs which is obſerved in part of Africa, and in Lapland. Climate enters there, in a great degree, for the effect; and it is aided by an abſurd ſenſe of beauty that prompts them often to depreſs it by art*.

The preceding obſervations tend to account for ſome of the moſt diſtinguiſhing features of ſavages. To theſe I might have added another general reaſon of their peculiar wildneſs and uncouthneſs in that ſtate of ſociety.—The feelings of ſavages, when they deviate from their uſual apathy, are moſtly of the uneaſy kind; and to theſe they give an unconſtrained expreſſion. From this cauſe will neceſſarily reſult a habit of the face, in the

we ſhall find a ſtricture on the noſe that, in an age when the features were ſoft and pliant, would ſenſibly tend to depreſs it. A like tendency continued through the whole of life, would give them an habitual poſition very different from the common condition of civilized ſociety; and the effect would be much greater than would readily occur to our firſt reflections upon the ſubject.

* That ſuch an effect ſhould be the reſult of climate is not more wonderful than the thick necks created by the climate of the Alps; or than other effects that certainly ſpring from this cauſe, within our own knowledge. That it ariſes from climate, or the ſtate of ſociety, or both, is evident, becauſe the noſe is becoming more prominent in the poſterity of thoſe who have been removed from Africa to America.

the higheſt degree rude and uncouth. As we ſee, a ſimilar negligence among the vulgar adds exceedingly, to that diſguſting coarſeneſs which ſo many other cauſes contribute to create.

I have now finiſhed the diſcuſſion which I propoſed, as far as I deſign at preſent to purſue it.—Many of the obſervations which have been made in the progreſs of it may, to perſons not accuſtomed to a nice examination of the powers of natural cauſes, appear minute and unimportant. It may be thought that I have attributed too much to the influence of principles that are ſo ſlow in their operation and imperceptible in their progreſs. But, on this ſubject, it deſerves to be remembered, that the minuteſt cauſes, by acting conſtantly, are often productive of the greateſt conſequences. The inceſſant drop wears a cavity, at length in the hardeſt rock. The impreſſions of education which ſingly taken are ſcarcely diſcernible, ultimately produce the greateſt differences between men in ſociety. How ſlow the progreſs of civilization which the influence of two thouſand years hath as yet, hardly ripened in the nations of Europe? How minute and imperceptible the operation of of each particular caufe that has contributed to the final refult? And, yet, how immenfe the difference between the manners of Europe barbarous, and of Europe civilized? There is furely not a greater difference between the figure and afpect of any two nations on the globe. The pliant nature of man is fufceptible of change from the minuteft caufes, and thefe changes, habitually repeated, create at length, confpicuous diftinctions. The effect proceeds increafing from one generation to another, till it arrives at that point where the conftitution can yield no farther to the power of the operating caufe. Here it affumes a permanent form and becomes the character of the climate or the nation.

Superficial thinkers are often heard to afk, why, unlefs there be an original difference in the fpecies of men, are not all *born* at leaft with the fame figure, or complexion? It is fufficient to anfwer to fuch enquiries, that it is for the fame reafon, whatever that may be, that other refemblances of parents are communicated to children. We fee that figure, ftature, complexion, features, difeafes, and even powers of the mind become hereditary. To thofe who can fatisfy themfelves with regard

gard to the communication of theſe properties, the tranſmiſſion of climatical or national differences ought not to appear ſurprizing—the ſame law will account for both.—If it be aſked why a ſun burnt face or a wounded limb is not alſo communicated by the ſame law? It is ſufficient to anſwer that theſe are only partial accidents which do not change the inward form and temperament of the conſtitution. It is the conſtitution that is conveyed by birth. The cauſes which I have attempted to illuſtrate, change, in time, its whole ſtructure and compoſition—And when any change becomes incorporated, ſo to ſpeak, it is, along with other conſtitutional properties, tranſmitted to offspring.

I proceed now to conſider the exceptions exiſting among mankind that ſeem to contradict the general principles that have been laid down concerning the influence of climate, and of the ſtate of ſociety.

I begin with obſerving that theſe exceptions are neither ſo numerous nor ſo great as they have been repreſented by ignorant and inaccurate travellers, and by credulous philoſophers. Even Buffon ſeems to be credu-

lous

lous when he only doubts concerning the relations of Struys, and other prodigy-mongers, who have filled the hiſtories of their voyages with crude and haſty obſervations, the effects of falſehood, or of ſtupid ſurprize. Nothing can appear more contemptible than philoſophers with ſolemn faces, retailing like maids and nurſes, the ſtories of giants*—of tailed men†—of a people without teeth‡—and of ſome abſolutely without necks§. It is a ſhame for philoſophy at this day to be ſwallowing the falſehoods, and accounting for the abſurdities of ſailors. We in America, perhaps, receive ſuch tales with more contempt than other nations; becauſe we perceive in ſuch a

N ſtrong

* Buffon, deſcribing the inhabitants of the Marian, or Ladrone iſlands, ſuppoſes that they are, in general, a people of large ſize; and that ſome may have been ſeen there of gigantic ſtature. But before Buffon wrote, there was hardly a navigator who did not ſee many giants in remote countries. Buffon has the merit of rejecting a great number of incredible narrations.

† Lord Monboddo ſuppoſes that mankind, at firſt, had tails—that they have fallen off by civilization—but that there are ſtill ſome nations, and ſome individuals who have this honorable mark of affinity with the brutes. What effect might reſult from the conjunction of a ſavage with an ape, or an Orang-Outang, it is impoſſible to ſay. But a monſtrous birth, if it ſhould happen, however it may be exaggerated by the ignorance of ſailors, ſhould never be dignified as a ſpecies in the writings of philoſophers.

‡ A moſt deformed and deteſtable people whom Buffon ſpeaks of as natives of New-Holland.

§ Sir Walter Raleigh pretends to deſcribe a people of that kind in Guiana. Other voyagers have given a ſimilar account of ſome of the Tartar tribes. The necks of theſe Tartars are naturally extremely ſhort; and the ſpirit of travelling prodigy has totally deſtroyed them.

ſtrong light, the falſehood of ſimilar wonders, with regard to this continent, that were a few years ago reported, believed, and philoſophiſed on in Europe. We hear every day the abſurd remarks, and the falſe reaſonings of foreigners on almoſt every object that comes under their obſervation in this new region. They judge of things, of men, and of manners under the influence of habits and ideas framed in a different climate, and a different ſtate of ſociety; or they infer general and erroneous concluſions from ſingle and miſtaken facts, viewed through that prejudice, which previous habits always form in common minds*.

Since

* It requires a greater portion of reflexion and philoſophy than falls to the lot of ordinary travellers to enable them to judge with propriety of men and things in diſtant countries. Countries are deſcribed from a ſingle ſpot, manners from a ſingle action, and men from the firſt man that is ſeen on a foreign ſhore, and perhaps him only half ſeen, and at a diſtance. From this ſpirit, America has been repreſented by different travellers as the moſt fertile or the moſt barren region on the globe. Navigators to Africa often ſpeak of the ſpreading foreſts and luxuriant herbage of that arid continent, becauſe ſome ſcenes of this kind are preſented to the eye along the ſhores of the Gambia and the Senegal. And ſurprize occaſioned by an uncommon complexion or compoſition of features, has increaſed or diminiſhed the ſtature of different nations beyond all the proportions of nature.—Such judgments are ſimilar, perhaps, to thoſe which a Chineſe ſailor would form of the United States who had ſeen only cape May; or would form of Britain or of France, who had ſeen only the ports of Dover or of Calais. What information concerning thoſe kingdoms could ſuch a viſitant afford his countrymen from ſuch a viſit? Beſide the limited ſphere of his obſervation, he would ſee every thing with aſtoniſhment or with diſguſt, that would exaggerate, or diſtort his repreſentation. He would ſee each action by itſelf without knowing its connexions; or he would ſee it with the connexions which it would have in his own country. A ſimilar

error

Since America is better known, we find no canibals in Florida; no men in Guiana with heads ſunk into their breaſts; no martial Amazons. The giants of Patagonia have diſappeared; and the ſame fate ſhould have attended thoſe of the Ladrone iſlands, whom Buffon after Gamelli Carreri has been pleaſed to mention. Tavernier's tales of the ſmooth and hairleſs bodies of the Mogul women, may be

error induced Capt. Cook in his firſt voyage, to form an unfavourable opinion of the modeſty and chaſtity of the women of Otaheite, which more experience taught him to correct. Many ſuch falſe judgments are to be found in almoſt every writer of voyages or travels. The ſavages of America are repreſented as frigid, becauſe they are not ready for ever to avail themſelves of the opportunities offered by their ſtate of ſociety, to violate the chaſtity of their females. They are ſometimes repreſented as licentious, becauſe they often lie promiſcuouſly round the ſame fire. Both judgments are falſe, and formed on prepoſſeſſions created in ſociety. Simplicity of manners, more than conſtitution, or than climate, produces that appearance of indifference, on the one hand, that is called frigidity, and that promiſcuous intercourſe, on the other, that is ſuppoſed to be united with licence. Luxury, reſtraints, and the arts of poliſhed ſociety inflame deſire, which is allayed by the coarſe manners, and hard fare of ſavage life, where no ſtudied excitements are uſed to awaken the paſſions. The frontier counties of all theſe ſtates at preſent afford a ſtriking example of the truth of this reflexion. Poor, and approaching the roughneſs and ſimplicity of ſavage manners, and living in cabins that have no diviſions of apartments, whole families, and frequently ſtrangers lodge together in the ſame incloſure without any ſenſe of indecency, and with fewer violations of chaſtity than are found amidſt the reſtraints and incitements of more poliſhed ſociety. On a like foundation cowardice has been imputed to the natives of America, becauſe they proſecute their wars by ſtratagem—inſenſibility, becauſe they ſuffer with patience—and thieviſhneſs, becauſe a ſavage, having no notion of perſonal property but that which he has in preſent occupation and enjoyment, takes without ſcruple what *he* wants, and ſees *you* do not need. In innumerable inſtances the act of one man, the figure or ſtature of the firſt vagrant ſeen upon a diſtant ſhore, has furniſhed the character of a whole nation. It is abſurd to build philoſophic theories on the ground of ſuch ſtories.

be ranked with thofe which have fo long, and fo falfely attributed this peculiarity to the natives of America. The fame judgment may we form of thofe hiftories which reprefent nations without natural affection; without ideas of religion; and without moral principle*. In a word, the greater part of thofe extraordinary deviations from the laws of climate, and of fociety, which formerly obtained credit, are difcovered, by more accurate obfervation, to have no exiftence. If a few marvellous phænomena are ftill retailed by credulous writers, a fhort time will explode them all, or fhew that they are mifunderftood, and enable

* Nations have been judged to be without religion becaufe travellers have not feen temples; becaufe they have not underftood their cuftoms, or their language; or have not feen them engaged in any act of worfhip. Nations have been judged to be without natural affection, becaufe one man has been feen to do an act of barbarity. But one of the nations which feems to have departed fartheft from the laws of human nature is mentioned by lord Kaims in his laudable attempts to difprove the truth of revelation. He thinks it certain that the Giagas, a nation of Africa, could not have defcended from one origin with the reft of mankind, becaufe, totally unlike all others, they are void of natural affection. They kill, fays his lordfhip, all their own children as foon as they are born, and fupply their places with youth ftolen from the neighbouring tribes. If this character had been true, even his lordfhip's zeal for a good caufe, might have fuffered him to reflect that the Giagas could not have continued a feparate race, longer than the firft ftock fhould have lived. The ftolen youth would refemble their parents, and would, at length, compofe the nation. And yet the Giagas, according to his lordfhip, would continue to kill their children, and to be a ftanding monument of the falfehood of the fcriptures! An excellent fpecimen of the eafy faith of infidelity!—Prelim. Difc. to Sketches of the Hift. of Man, by lord Kaims.

able philosophers to explain them on the known principles of human nature.

Leaving such pretended facts and the reasonings to which they have given rise, to deserved contempt, I shall consider a few apparent deviations from the preceding principles that have been ascertained. It will not be necessary to go into an extensive detail of minute differences. These might be tedious and unimportant: I shall propose only the most conspicuous, persuaded that, if they are satisfactorily explained, every reasonable inquirer will rest convinced that natural causes exist in every country sufficient to account for smaller distinctions.

In tracing the same parallels from east to west, we do not always discern the same features and complexion. In the countries of India, and on the northern coasts of Africa, nations are mingled together who are distinguished from one another by great varieties. The torrid zone of Asia is not marked by such a deep colour nor by such parched hair as that of Africa; and the colour of tropical America is, in general, lighter, than that of Asia.

Africa is not uniform. The complexion of the weſtern coaſt is a deeper black than that of the eaſtern. It is even deeper on the north of the equator than on the ſouth. The Abyſſinians form an exception from all the other inhabitants of the African zone—and when we go beyond that zone to the ſouth, the Hottentots ſeem to be a race by themſelves. In their manners the moſt beaſtly, and in their perſons and the faculties of their minds approaching the neareſt to brutes of any of the human ſpecies.

For the explication of theſe varieties it is neceſſary to obſerve that the ſame parallel of latitude does not uniformly indicate the ſame temperature of heat and cold. Vicinity to the ſea, the courſe of winds, the altitude of lands, and even the nature of the ſoil, create great differences in the ſame climate. The ſtate of ſociety in which any nation takes poſſeſſion of a new country has a great effect in preſerving, or in changing their original appearance. Savages neceſſarily undergo great changes by ſuffering the whole action and force of climate without protection. Men in a civilized ſtate enjoy innumerable arts by which they are enabled to guard againſt its influence,

influence, and to retain ſome favourite idea of beauty formed in their primitive ſeats. Yet, every migration produces a change. And the combined effects of many migrations, ſuch as have been made by almoſt all the preſent nations of the temperate zone, muſt have great influence in varying the human countenance. For example—A nation which migrates to a different climate will, in time, be impreſſed with the characters of its new ſtate. If this nation ſhould afterwards return to its original *ſeats*, it would not perfectly recover its primitive features and complexion, but would receive the impreſſions of the firſt climate, on the ground of thoſe created in the ſecond. In a new removal the combined effect of the two climates, would become the ground, on which would be impreſſed, the characters of the third. This exhibits a new cauſe of endleſs variety in the human countenance.

Theſe principles will ſerve to explain many of the differences that exiſt in thoſe countries which have been the ſubjects of moſt frequent conqueſt*. India and the northern regions

* Eſpecially if religion, manners, policy, or other cauſes, prevent people from uniting freely in marriages, and from ſubmitting to the ſame ſyſtem of government and laws.

regions of Africa, have been often conquered, and many nations have eſtabliſhed colonies in theſe countries for the purpoſes of commerce. All theſe nations before their migrations, or their conqueſts, were in a leſs or greater degree, civilized. They were able therefore, to preſerve, with ſome ſucceſs their original features againſt the influence of the climate. Their diet, their habits, their manners and their arts, all would contribute to this effect. As theſe cauſes are capable of creating great varieties among men, much more are they capable of preſerving varieties already created. The Turks therefore, the Arabs, and the Moors in the north of Africa, will remain, forever, diſtinct in their figure and complexion, as long as their manners are different. And the continent and iſlands of India will be filled with a various race of people while the productions of their climate continue to invite both conqueſts and commerce. The climate will certainly change in a degree the appearance of all the nations who remove thither; but the difference in the degree and the combination of this effect with their original characters, will ſtill preſerve among them eſſential and conſpicuous diſtinctions*.

Another

* From the preceding principles we may juſtly conclude that the Anglo-Americans

Another variety which ſeems to form an exception from the principles hitherto laid down; but which really eſtabliſhes them, is that the torrid zone of Aſia is not marked by ſuch a deep colour, nor, except in a few countries, by ſuch curled hair, as that of Africa. The African zone is a region of burning ſand which augments the heats of the ſun to a degree almoſt inconceivable. That of Aſia, conſiſts chiefly of water which, abſorbing the rays of the ſun, and filling the atmoſphere with a cool and humid vapour, creates a wind comparatively temperate over its numerous iſlands and narrow peninſulas. The principle body of its lands lies nearer to the northern tropic than to the equator. In ſummer the winds blow from the ſouth acroſs extenſive oceans; in the winter from conti-

O nents

Americans will never reſemble the native Indians. Their civilization will prevent ſo great a degeneracy. But were it poſſible that they ſhould become ſavage, the reſemblance could never be complete, becauſe the one would receive the impreſſions of the climate on a countenance, the ground of which was formed in Europe, and in a ſtate of improved ſociety; the other has plainly received them on a countenance formed in Tartary. And yet the reſemblance becomes near and ſtriking in thoſe perſons who have been captivated by the Indians in infancy, and have grown up among them in the habits of ſavage life. Theſe principles likewiſe will lead us to conclude that the Samoiedes are Tartars degenerated by the effects of extreme cold—and that the empire of China and moſt of the countries of India have been peopled from the north. For their countenance ſeems to be compoſed of the ſoft feature of the Lower Aſia, laid upon a ground formed in the Upper Aſia.

nents that the ſun has long deſerted*. Yet, under all the advantages of climate which Aſia enjoys, we find in Borneo and New-Guinea, and perhaps in ſome others of thoſe vaſt inſular countries, which, by their poſition and extent, are ſubject to greater heats than the continent, or by the ſavage condition of the inhabitants, ſuffer the influence of thoſe heats, in a higher degree, a race of men reſembling the African negroes. Their hair, their complexion and their features, are nearly the ſame. At the diſtance of more than three thouſand miles acroſs the Indian ocean, it is impoſſible that they ſhould have ſprung from the ſavages of Africa, who have not the means of making ſuch extenſive voyages†. Similarity of climate, and of manners, have created this ſtriking reſemblance, between people ſo remote from one another.

The next apparent exception, we diſcover in Africa itſelf. Africa, like Europe and Aſia, is full of varieties, ariſing from the ſame cauſes, vicinity to the ſun, elevation of the land, the heat of winds, and the manners of the

* The monſoons are found to blow over the whole Aſiatic zone.

† The Europeans were highly civilized before they diſcovered the continent of America, which is not ſo remote from their ſhores as Borneo or New-Holland is from the coaſt of Africa.

the people. But the two principal diſtinctions of colour, under which the reſt may be ranged, that prevail from the northern tropic, or a little higher to the cape of Good-Hope, are the Caffre and the negroe. The Caffre complexion prevails along the eaſtern coaſt, and in the country of the Hottentots. The negroe, on the weſtern coaſt between the tropics. The negroe is the blackeſt colour of the human ſkin, the Caffre is much lighter and ſeems to be the intermediate grade between the negroe and the native of India. The cauſe of this difference will be obvious to thoſe who are acquainted with that continent. The winds under the equator, following the courſe of the ſun, reach the eaſtern coaſt of Africa, cooled by blowing over immenſe oceans, and render the countries of Aian, Zanguebar and Monomotapa, comparatively temperate. But after they have traverſed that extenſive continent, and in a paſſage of three thouſand miles have collected all the fires of the burning deſert to pour them on the countries of Guinea, of Sierra-Leona, and of Senegal*, they glow with an ardor

* Theſe countries receive the wind after blowing over the wideſt and hotteſt part of Africa, and conſequently ſuffer under a more intenſe heat than the countries of Congo, Angola, or Loango to the ſouth of the equator. Accordingly, we find the people of a deeper black in the northern than in the ſouthern ſection of the torrid zone.

ardor unknown in any other portion of the globe. The intenſe heat, which, in this region, makes ſuch a prodigious change on the human conſtitution, equally transforms the whole race of beaſts and of vegetables. All nature bears the marks of a powerful fire*. And the negroe is no more changed from the Caffre, the Moor, or the European, than the proportional laws of climate, and of ſociety give us reaſon to expect. Above the Senegal we find in the nation of the Foulies a lighter ſhade of the negroe colour; and immediately beyond them to the north, the darkeſt copper of the Mooriſh complexion. There is a ſmaller interval between the copper colour and the perfectly black on the north than on the ſouth of the torrid zone; becauſe the Moors being more civilized than the Hottentots are better able to defend themſelves againſt the impreſſions of the climate. But the Hottentots, being the moſt ſavage of mankind, ſuffer the influence of their climate in

* The luxuriancy of the trees and herbage along the banks of the great rivers has deceived ſome travellers who have repreſented Africa as a rich and fertile country. As ſoon as you leave the rivers, which are very few, you enter on a parched and naked ſoil. And the whole interior parts of that continent as far as they have been explored, are little elſe than a deſert of burning ſand, that often rolls in waves like the ocean. Buffon mentions a nation in the center of Africa, the Zuinges, who, the Arabian writers ſay, are often almoſt intirely cut off, by hot winds that riſe out of the ſurrounding deſerts.

in the extreme. And they endeavour, by every mean to preſerve the features and the complexion of the equator, from whence, it is probable, they derived, with their anceſtors, their ideas of beauty. It is more eaſy to preſerve acquired features or complexion, than to regain them after they have been loſt. The Hottentots preſerve with ſome ſucceſs, thoſe that they had acquired under the equator. They flatten, by violence, the noſe of every child ſoon after it is born; they endeavour to deepen the colour of the ſkin by rubbing it with the moſt filthy unguents, and by expoſing it to the influence of a ſcorching ſun; and their hair they burn up by the vileſt compoſitions. Yet, againſt all their efforts, the climate, though it is but a few degrees declined from the torrid zone, viſibly prevails. Their hair is thicker and longer than that of the negroes; and their complexion near the Cape is the lighteſt ſtain of the Caffre colour. Allowing for the effects of their ſavage condition, and of their brutal manners, they are marked nearly with the ſame hue that diſtinguiſhes the correſpondent northern latitudes*.

As

* With regard to other peculiarities that have been related of this people, and that reduce them in their figure the neareſt to the brute creation of any of the human ſpecies, great part of them are falſe, others exaggerated, and thoſe that are true are the natural offspring of their brutal manners.

As you aſcend along the eaſtern coaſt from Cafraria to Aian, the complexion becomes gradually deeper, till ſuddenly you find, in Abyſſinia, a race of men reſembling the ſouthern Arabians. Their hair is long and ſtraight, their features tolerably regular, and their complexion a very dark olive approaching to the black. This ſingularity is eaſily explained on the principles already eſtabliſhed: and it is an additional confirmation of theſe principles that they are found to reach all the effects to which they are applied. The Abyſſinians are a civilized people, and bear evident marks of Aſiatic origin. They are ſituated in the mildeſt region of tropical Africa, and are fanned by the temperate winds that blow from the Indian ocean. Abyſſinia is likewiſe a high and mountainous country, and is waſhed during half the year by deluges of rain which impart unuſual coolneſs to the air. It is, perhaps, one of the moſt elevated regions on earth, as, from its mountains ſpring two of the largeſt and the longeſt rivers in the world, the Niger and the Nile*. This altitude

* The prodigious and inceſſant deluges of rain that fall in Abyſſinia during ſix months in the year, are the cauſe of the overflowing of the Nile. They render the atmoſphere temperate, and are a proof of the elevation of the country, no leſs than the length of the rivers that originate in its mountains.

tude of the lands, raiſes it to a region of the atmoſphere that is equivalent to many degrees of northern latitude*. Thus, the civilization of the people, the elevation of the country, the temperature of the winds, and inceſſant clouds and rain during that ſeaſon of the year in which the ſun is vertical, all contribute to create that form and colour of the human perſon in Abyſſinia, which is conſidered as a prodigy in the torrid zone of Africa.

Having conſidered the principal objections to the preceding theory exiſting in India and Africa, it may be expected that I ſhould not omit to mention the white Negroes of Africa, and the white Indians of Darien, and of ſome of the oriental iſlands, which are ſo often quoted upon this ſubject. Ignorant or intereſted writers have endeavoured to magnify this phænomenon into an argument for the original diſtinction of ſpecies. But thoſe who have

mountains. The greateſt quantity of rains uſually fall on mountains and the higheſt lands; and their elevation may, in a great meaſure, be determined by the length of the rivers that iſſue from them.

* Some writers inform us that the barometer riſes in Abyſſinia, on an average, no higher than 20 inches. If this be true, that kingdom muſt be ſituated more than two miles above the level of the ſea. But if we ſhould ſuppoſe this account to be exaggerated, ſtill we muſt judge its altitude to be very great, conſidering that it is almoſt intirely a region of mountains, which are the ſources of thoſe vaſt rivers.

have examined the fact with greater accuracy, have rendered it evident that their colour is the effect of some distemper. These whites are rare; they have all the marks of an extreme imbecility; they do not form a separate race, or continue their own species; but are found to be the accidental and diseased production of parents who themselves possess the full characters of the climate*.

It now remains only to account for the aspect of the savage natives of America, which varies from the examples we have considered in the other portions of the earth. Their complexion is not so fair as that of Europe or of Middle Asia. It is not so black as that of Africa,

* Mr. James Lind, a physician of great reputation, has recorded a similar deviation from the law of climate in a black child born of white parents. The fact he assures us occurred to his own observation. See Phil. Trans. of Roy. Soc. Lond. N° 424.

The small tribe of red people, which Dr. Shaw, in his travels, relates that he saw in the mountains of Auress, a part of the vast ridge of Atlas, are probably a remnant of the Vandals who, in the fifth century, conquered the northern countries of Africa. Their manners, and the altitude of their situation, in those cold mountains, may have contributed to preserve this distinction between them and the Moors and Arabs, who live in the low lands. Lord Kaims, who writes with infinite weakness on this subject, exclaims with an air of triumph, if the climate in a thousand years has not changed these people into a perfect resemblance of the aborigines, we may safely pronounce it never will change them,—I confess it, if they preserve their present elevation. But to conclude that the climate cannot change them on the plains, because it has not changed them on the mountains, is the same kind of reasoning as it would be to conclude that the sun could not melt snow at the bottom of Ætna or Pambamarca because it continues eternally frozen at the top.

Africa, and many of the oriental iſlands. There is a greater uniformity of countenance throughout this whole continent than is found in any other region of the globe of equal extent.

That the natives of America are not fair, is a natural conſequence of the principles already eſtabliſhed in this eſſay; in which it has been ſhewn that ſavages, from their expoſure, their hardſhips, and their manner of living, muſt, even in temperate climates, be diſcoloured by different ſhades of the tawny complexion.

The uniformity of their countenance reſults in ſome degree from that of the climate, which is the leſs various, that America poſſeſſes the cooleſt tropical region in the world. But it reſults principally from their ſtate of ſociety, their manners, their means of ſubſiſtence, the nature and limitation of their ideas, which preſerve an uncommon reſemblance from Canada to cape Horn. Though complexion is leſs diverſified in America than in other regions of the earth; yet there is a ſenſible gradation of colour*, till you arrive at the darkeſt

* In travelling from the great lakes to Florida or Louiſiana, through the Indian nations, there is a viſible progreſſion in the darkneſs of their complexion.

hue of this continent in the nations on the west of Brazil. Here the continent being wider, and consequently hotter, than in any other part between the tropics, is more deeply coloured. And the Toupinamboes and Toupayas, and other tribes of that region, bear a near resemblance, in their complexion, to the inhabitants of the oriental zone. We find indeed no people in America so black as the Africans. This is the peculiarity that attracts most observation and inquiry; and the cause, I propose now to explain.

The torrid zone of America is uncommonly temperate. This effect arises in part from its shape; in part from its high mountains, and extensive lakes and rivers; and in part from its uncultivated state. All uncultivated regions, covered with forests and with waters, are naturally cold*. The torrid zone of America

plexion. And at the councils of confederate nations, or at treaties for terminating an extensive war, you often see sachems and warriors of very different hues. But the colour of the natives of America, though diversified, is less various than in other quarters of the globe of equal extent of latitude. And as the same state of society universally prevails, there is a system of features that results from this, which is every where similar. These features giving the predominant aspect to the face, and being united with a complexion less various than in Africa or Asia, form what is called the uniformity of the American countenance.

* The difference, in point of climate, which cultivation has produced between modern and ancient Europe, is well known. And it is probable that,

rica is narrow—its mountains and its rivers are immenſe—and Amazonia may be conſidered, during a great portion of the year, as one extenſive lake†. Let us advert to the influence of theſe circumſtances. The empire of Mexico is a continued iſthmus of high and mountainous lands. Cool by their elevation, they are fanned on each ſide by winds from the eaſtern and weſtern oceans. Terra Firma is a hilly region. Amazonia, though low and flat, is ſhaded by boundleſs foreſts, and cooled by the numerous waters that flow into the largeſt rivers in the world. The mildneſs of its atmoſphere is augmented by the perpetual eaſt wind that blows under the equator. This wind having depoſited in the Atlantic ocean the heats acquired in its paſſage acroſs the continent of Africa, regains a moderate temperature before it arrives at the American coaſt. In America it continues its courſe over thick foreſts and innumerable waters, to the mountains of the Andes. The Andes are colder than the Alps. And the empire of Peru defended, on one ſide, by theſe frozen ridges;

that, if civilization ſhall, in future time, be introduced into Tartary, that frozen climate will be mollified, and the deformed Tartars may, with change of climate and of manners, become perſonable men.

† On account of its numerous rivers and its flooded lands.

ridges; fanned on the other by a perpetual weſt wind from the Pacific ocean; and covered by a canopy of denſe vapour, through which the ſun never penetrates with force, enjoys a temperate atmoſphere. The vaſt foreſts of America are an effect of the temperature of the air, and contribute to promote it. Extreme heat parches the ſoil, and converts it into an arid ſand—luxuriant vegetation is the fruit of a moiſt earth, and a temperate ſky. And the natives, inhabiting perpetual ſhade, and reſpiring in the grateful and refrigerating effluvia of vegetables, enjoy, in the midſt of the torrid zone, a moderate climate.

Theſe obſervations tend to ſhew that, as far as heat is concerned in the effect, the colour of the American muſt be much leſs deep than that of the African, or even of the Aſiatic zone. And to me it appears, and, I doubt not, to every candid and intelligent inquirer, that the co-operation of ſo many cauſes is fully adequate to account for the differences between the complexion of the Negroe, and of the Indian.

Thus

Thus have I concluded the examination, which I propofed, into the caufes of the principle varieties of perfon that appear in the different nations of the earth. And I am happy to obferve, on this fubject, that the moft accurate inveftigations into the power of nature ever ferve to confirm the facts vouched by the authority of revelation. A juft philofophy will always be found to be coincident with true theology. The writers who, through ignorance of nature, or through prejudice againft religion, attempt to deny the unity of the human fpecies do not advert to the confufion which fuch principles tend to introduce. The fcience of morals would be abfurd; the law of nature and nations would be annihilated; no general principles of human conduct, of religion, or of policy could be framed; for, human nature, originally, infinitely various, and, by the changes of the world, infinitely mixed, could not be comprehended in any fyftem. The rules which would refult from the ftudy of our own nature, would not apply to the natives of other countries who would be of different fpecies; perhaps, not to two families in our own country, who might be fprung from a diffimilar compofition of fpecies. Such principles tend to confound all ſcience, as well as piety; and leave us in the world uncertain whom to truſt, or what opinions to frame of others. The doctrine of one race, removes this uncertainty, renders human nature ſuſceptible of ſyſtem, illuſtrates the powers of phyſical cauſes, and opens a rich and extenſive field for moral ſcience. The unity of the human race I have confirmed by explaining the cauſes of its variety.—The firſt and chief of theſe I have ſhewn to be climate; by which is meant, not ſo much the latitude of a country from the equator, as the degree of heat or cold, that depends on many connected circumſtances. The next, is the ſtate of ſociety, which greatly augments or corrects the influence of climate, and is itſelf the independent cauſe of many conſpicuous diſtinctions among mankind. Theſe cauſes may be infinitely varied in their degree, and in their combinations with other principles. And in the innumerable migrations of mankind, they are modified by their own previous effects in a prior climate, and a prior ſtate of ſociety*. Even where all external circumſtances ſeem to be the ſame, there may be ſecret cauſes of difference, as there are varieties in the children of

* Vide pages, 95 and 96.

of the ſame family. The ſame country often exhibits differences among individuals ſimilar to thoſe which diſtinguiſh the moſt diſtant nations. Such differences prove, at leaſt, that the human conſtitution is ſuſceptible of all the changes that are ſeen among men. It is not more aſtoniſhing that nations, than that individuals ſhould differ †. In the one caſe, we know with certainty, that the varieties have ariſen out of the ſame origin; and, in the other, we have reaſon to conclude, independently on the ſacred authority of revelation, that from one pair have ſprung all the families of the earth.

† It would be lawful, if it were neceſſary, to have recourſe to accidental cauſes to account for the varieties of nations; and to ſuppoſe that a country might have, at firſt, been peopled by ſome anceſtor moſt like the natives in features and in figure. It would not be a ſtrained ſuppoſition, becauſe we frequently ſee deformed perſons in civil ſociety reſemble almoſt every ſavage nation. And thoſe who are acquainted with American migrations know, that, commonly, the moſt poor, and lazy, and deformed, are the firſt to puſh their fortune in a rude and ſavage wilderneſs, where they can live, without labour, by fiſhing and hunting.

F I N I S.

STRICTURES

ON

LORD KAIMS's DISCOURSE

ON THE

ORIGINAL DIVERSITY OF MANKIND.

LORD Kaims, in a preliminary diſcourſe to his ſketches of the hiſtory of man, has undertaken to combat the principle which I have endeavoured to maintain, that all mankind are ſprung from one pair. His reputation ſtands ſo high in the literary world, that we may juſtly preſume he has comprehended in that diſſertation whatever can be urged with ſolidity againſt this opinion. Every reader will probably deem the refutation of ſuch an antagoniſt, no inconſiderable addition to the force of the preceding argument.

The character of lord Kaims, as an author, appears in this diſcourſe, far inferior to that which he has juſtly obtained from his other works. And in ſome ſtrictures which I am now to make upon it, I propoſe to ſhew that many of the ſuppoſed facts on which his lordſhip relies in the train of his argument, have no exiſtence, and that almoſt the whole of his reaſoning is inconcluſive.

In the firſt place he ſays, " certain it is that all " men, more than all animals, are not equally fitted " for every climate. There were therefore created " different kinds of men at firſt, according to the na- " ture of the climate in which they were to live. " And if we have any belief in providence, it ought " to be ſo. Becauſe men, in changing their climate " uſually become ſickly and often degenerate."

This power of the climate to *change the perſon* which his lordſhip confeſſes, when he calls it the *degenerating* of mankind, is the principle for which I plead; and which, united with the influence of the ſtate of ſociety, is ſufficient to explain all the changes that are viſible in the different nations of the earth. Are not the inhabitants of Guinea and of Lapland, degenerated races compared with the inhabitants of France and England? If theſe people had, in their own climates, attained the perfection of their nature, and the civilized Europeans had, by being tranſplanted thither, degenerated far below them, the argument then would have had ſome force. But ſince the greateſt degeneracy of Europeans is only a reſemblance of theſe ſavages, the example concludes againſt his lordſhip's principle.

But men, he contends were not made for different climates, " becauſe, in changing their climate, they " uſually become ſickly."

This argument ſuppoſes that man was not made for ſituations in which he is liable to encounter dan-

ger

ger or diſeaſe. And yet we ſee him, as it were by the appointment of providence, continually encountering both. If this argument were of weight, man is only an intruder on this world; for, every where he meets with ſickneſs, and with death. True it is, men, by making great and ſudden changes of climate or of country, are expoſed to diſeaſe. But it is equally true of ſimilar changes even in the modes of living. And the argument proves only that all ſuch alterations ſhould be made gradually, and with precaution. If this prudential conduct be obſerved, the human conſtitution, as is known from actual experiment, is capable of enduring the influence of every climate. It becomes, in time, aſſimilated by its ſituation. And the progeny of foreigners come at length to reſemble the natives, if they adopt the ſame manners.---In America we are liable to diſorder, by removing incautiouſly from a northern to a ſouthern ſtate; and even from one part to another of the ſame ſtate: but it would be abſurd to conclude thence, that we are not of one ſpecies from New-Hampſhire to Georgia. Shall we conclude that the top of every hill, and the bank of every river are inhabited by different ſpecies, becauſe the latter are leſs healthy than the former? The conſtitution becomes attempered, in a degree even to an unhealthy region, and then it feels augmented ſymptoms of diſorder, on returning to the moſt ſalubrious air and water: but does this prove that nature never intended ſuch men to drink clear water, or to breathe in a pure atmoſphere? This argument deſtroys itſelf by the extent of the conſequences which it draws after it.

His

His lordſhip's ſecond argument which is only a repetition of part of the firſt, is certainly an extraordinary example of philoſophic reaſoning---" Men, ſays " he, muſt have been originally of different ſtocks, " adapted to their reſpective climates, becauſe an Eu- " ropean degenerates both in vigour and in colour " on being removed to ſouth America, to Africa, or " to the Eaſt Indies."

The fact is as his lordſhip ſtates it. An European changes his colour on being removed to theſe diſtant climates. But one would think that true philoſophy ſhould have drawn from this fact a contrary concluſion. Certainly if an European had *not degenerated*, as he expreſſes it, in colour and in vigour, on being removed to other climates, it would have been a ſtronger proof of the original difference of races.

He confirms this obſervation, however, by the example of " a Portugueſe colony on the coaſt of Con- " go, who in a courſe of time, he affirms, have de- " generated ſo much, that they ſcarce retain the ap- " pearance of men."

A fact more to the purpoſe of the preceding eſſay could not be adduced. Let it be applied to the neighbouring tribes of negroes and of Hottentots. Though they, in like manner, are become ſo rude that ſcarcely do they retain the appearance of men, does not his lordſhip's example prove that, in ſome remote period, they might have deſcended from the ſame origin with theſe degenerated Portugueſe?

His lordſhip has been egregiouſly deceived in the principle on which he attempts to prove that America is not adapted to European conſtitutions. He aſſerts that " Charleſtown in Carolina is inſufferably " hot; becauſe ſays he, it has no ſea-breeze---that " Jamaica itſelf is a more temperate climate--- and " that the inhabitants of both die ſo faſt that if con" tinual recruits did not arrive from Europe to ſup" ply the places of thoſe that periſh the countries " would be ſoon depopulated."---How cautious ſhould philoſophers be of aſſerting facts, without well examining the authority on which they receive them! All theſe aſſertions are equally and entirely falſe. And if a philoſopher, and a lord of ſeſſions in Scotland, talks ſo ignorantly of that country which, from its long and intimate connexion with Britain, he ſhould have underſtood better than any other, we may juſtly preſume that he is leſs acquainted with the Aſiatic and African nations; and that the objections drawn from them by him, and by inferior writers, againſt the doctrine of one race, are ſtill more weak and unfounded.

His lordſhip uſes, as another argument for the original diverſity of ſpecies among mankind that common European miſtake, that, " the natives of Ame" rica are deſtitute of hair on the chin and body."

That philoſophers ſhould ſometimes be deceived in their information is not ſurprizing; but they are certainly blameable, after having found in ſo many repeated examples the falſehood of voyagers, or their incapacity

incapacity for obſervation, to reſt, on ſuch dubious tales, an argument againſt the moſt common and ſacred opinions of mankind*.

His lordſhip, in the next place ſays with truth, that " the northern nations, to protect them from the cold, have more fat " than the ſouthern."---But from this principle he draws a falſe concluſion, that " therefore the northern and ſouthern nations are of different races, adapted by nature to their reſpective climates." ---He ought to have drawn the contrary concluſion, that nature hath given ſuch pliancy to the human conſtitution as to enable it to adapt itſelf to every clime. The goodneſs of the Creator appears in forming the whole world for man, and not confining *him*, like the inferior animals, to a bounded range, beyond which he cannot paſs either for the acquiſition of ſcience, or, for the enlargement of his habitation. And the divine wiſdom is ſeen in mingling in the human frame ſuch principles as always tend to counteract the hazards of a new ſituation. Fat protects the vitals from the too piercing influence of cold*. But this covering being too warm for ſouthern regions, nature hath enabled the conſtitution to throw it off by perſpiration. The

* I have ſhewn in the eſſay that this peculiarity has been falſely imputed to the natives of America; and that they are not, in this reſpect, diſtinguiſhed, by nature, from the reſt of mankind. They have a cuſtom, founded on a capricious idea of beauty, of pulling out their hair with tweezers. And haſty and ſuperficial travellers have been deceived, by the apparent ſmoothneſs of the chin and body, into the imagination, that they are naturally deſtitute of this excreſcence.

† Almoſt all animals that run wild in the foreſt, grow fatter at the approach of winter; and they ſtill augment their fat by being removed to a latitude farther north.

The phyſical cauſe of this effect ought to have been no ſecret to a philoſopher who treats of *human nature*. Not to mention the natural effects of the relaxation of heat; or the bracing of cold, on the nouriſhment of the body; it is ſufficient to obſerve, that the profuſe perſpiration that takes place in ſouthern latitudes, carries off the oily with the aqueous parts, and renders the conſtitution thin; but a frigid climate, by obſtructing the evaporation of the oils condenſes them in a coat of fat that contributes to preſerve the warmth of the animal ſyſtem. Experience verifies this influence of climate. The northern tribes which, iſſued from the foreſts of Germany, and overrun the ſouthern provinces of the Roman Empire, no longer retain their original groſſneſs, and their vaſt ſize. The conſtitution of Spain, and of other countries in the ſouth of Europe is thin; and the Europeans in general have become more thin by emigrating to America. Here is a double experiment, within the memory of hiſtory, made on intire nations. Many ſingle examples will occur to every man's obſervation. The argument, therefore, which this writer derives from the fatneſs of one nation, and the leanneſs of another is inconcluſive for the purpoſe for which he urges it, the proof of different ſpecies of men.

His next attempt is to prove that negroes are of a different ſpecies from whites. He ſays, "their ſkin "is more cool and adapted to their fervid climate. "For a thermometer applied to the body of an African, will not indicate the ſame degree of heat as "when applied to the body of an European."

The

The fact I will not dispute. But admitting it to be true with regard to the Europeans who travel to Africa, it is capable of explanation on the known principles of natural science. Perspiration from the human body is analogous to the evaporation of fluids, which is one of the most cooling processes in nature. It becomes a conductor to the internal heat, which it carries off as fast as it is excited, and thereby preserves the body in a moderate temperature. But when perspiration is obstructed, the retained heat immediately raises a fever in the system. The more profuse therefore the perspiration is, under the same degree of external heat, the more temperate will be the warmth of the skin*. In sweating, the skin is sensibly cooler than before the sweat begins to issue from the pores. In the torrid zone the heat relaxing and opening the pores of the natives will render both sensible and insensible perspiration in them more copious and constant, than in the natives of northern regions who remove thither. Their constitution not being yet perfectly accommodated to the climate, they do not perspire so freely. Being more full of blood, and highly toned, they suffer, in that fervid climate, the additional heat of an habitual fever. If the fact however be, as his lordship states it, the experiment must have been made on the whites in Africa, before the constitution

* For a similar reason likewise among others the surfaces of all fluids, preserve a greater coolness under the action of the direct rays of the sun than the surfaces of solid bodies. The action of the sun produces evaporation; and by this vapour the excited heat is conducted off, which, by remaining in solid bodies, renders them warmer than fluids. And this is equally true, whether we consider heat, with modern philosophers as an element, or with the old philosophers as only an internal commotion of parts.

constitution was properly reduced to suffer the intense heats of that region. For, in this climate, I can affirm from actual experiment, that the skin of a negroe is not cooler than that of a white person. I have applied the thermometer successively to two persons in my family of the same sex, and nearly of the same age, the one white, and the other black; and after making the trial in all respects as equal as possible, I have not been able at the end of half an hour to discover any difference in the elevation of the mercury.

Some of his lordship's following remarks and reasonings, I beg leave to treat a little more briefly.

" Is it possible, he asks, to account for the low stature, and little feet, and large head of the Esquimaux? or for the low stature and ugly visage of the Laplanders, by the action of cold?"

I have endeavoured to account for them from the *action of cold* in conjunction with *the state of society*.

" But the difference of latitude, he says, between the Laplanders, and the Norwegians and Fins, is not sufficient to account for the difference of features."

I have already explained the reason of this phænomenon. The temperate climates border upon eternal cold, and civilized on savage society, in every quarter of the globe. I have shewn that the forces of theſe two powerful cauſes combined, are fully adequate to account for theſe different effects.

His lordſhip confeſſes, that "it has been lately diſ-"covered by the *Pere Hel,* an Hungarian, that the "Laplanders were originally Huns."

Pere Hel has no doubt given authentic evidence of the fact, as appears by the conviction it has produced in his lordſhip. But it is ſtrange that it ſhould not have occurred to this ingenious writer, that, from the ſame Huns are deſcended, likewiſe, ſome of the moſt beautiful nations in Europe.

As an objection againſt the power of climate to change the complexion, he ſays, "the Moguls and "the ſouthern Chineſe are white." If he means that they are not black, it is true: If he means that they are as white as the Europeans, it is falſe. If the Moguls are leſs diſcoloured than ſome other nations in the ſame latitude, I have before aſſigned the reaſon. The ſtate of civilization to which they had arrived, previouſly to their taking poſſeſſion of their preſent ſeats, enabled them to defend themſelves with ſome ſucceſs againſt the impreſſions of a new climate.

His lordſhip adds, "Zaara is as hot as Guinea, "and Abyſſinia is hotter than Monomotapa, and yet "the inhabitants of the former are not ſo black as "thoſe of the latter." His lordſhip's hiſtorical, as well as phyſical knowledge, needs a little emendation. Zaara is not ſo hot as Guinea, nor is Abyſſinia ſo hot

as

as Monomotapa. But if it were equally hot, there are other causes that produce a wide difference between the figure and complexion of those nations*. The Abyssinians are civilized, the Monomotapans are savage. The Abyssinians derive their origin from Arabia; and civilization enables them to preserve their original features. The Monomotapans are evidently descended from the negroes of the equator, and their savage habits have continued the figure of their ancestors with little variation.

His lordship proceeds, "there are many instances "of races of people preserving their original colour "in climates very different from their own." This is nearly true of civilized nations, the reasons of which have been already assigned. It is not, however, by any means true, in the extent in which he asserts it†. He adds, "and there is not a single instance to the "contrary." To his lordship, the Portuguese of Congo might have been that instance.

Another argument for the original diversity of nations, on which some reliance is placed in this preliminary discourse, is taken from the variety of disposition, spirit and genius existing in different countries.

On this part of the subject some of his remarks are so ridiculously weak, that it is difficult to treat them with a serious face. Some of the oriental islands he mentions *whose inhabitants are hostile, and others whose* *inhabitants*

* See page 102 of the Essay.

† This has been sufficiently shewn in the precedng essay.

inhabitants are hoſpitable to ſtrangers, and thence concludes *a diverſity of ſpecies*. Kindneſs or averſion to ſtrangers depends on ſo many contingent cauſes that there cannot be a more equivocal foundation on which to reſt the argument for different races. Nations that have been often expoſed to hoſtile attacks, will be ſuſpicious of foreigners, and prone to repel them. Nations who have ſeldom ſeen the face of an enemy will be diſpoſed to receive them with kindneſs and hoſpitality. As well might he have proved, that Europe in the tenth, and in the eighteenth century, was inhabited by different ſpecies of men, from the facility and ſecurity with which a ſtranger can now paſs through all its kingdoms, and the hazards to which he was then expoſed. His lordſhip goes on to confirm this argument by the example of ſome nations *who are full of courage and prompt to combat*; and of others who hardly know the *arts of war*, or have *confidence to meet an enemy in battle*. With equal reaſon I might conclude that the Geeeks are not the ſame ſpecies now as when they gave birth to Ageſilaus, Miltiades, and Alexander. That the Romans were not the ſame ſpecies under Cæſar when they conquered, as under Auguſtulus when they loſt a world. And that, among the Jews, the Eſſenes, who were peaceful hermits in the foreſt, were not the ſame ſpecies with the martial Phariſees who reſiſted Titus. But the argument is too abſurd to merit even this anſwer.

He ſpeaks in the next place of the "*cowardice of the American Indians*," of whom he is manifeſtly ignorant,

ignorant, as a criterion of a distinct species. He proves the character, because they do not fight like the Europeans in an open field. An Indian philosopher, who should have examined the subject as superficially as lord Kaims, would probably retort the charge of cowardice on the Europeans, because they do not suffer torture like the natives of America. Nations have different ideas of courage and honour, and they exert these principles in different ways. The military education of an Indian consists in learning to make war by stealth, and to suffer with heroic fortitude. The reasons of their conduct in both, arise naturally out of their state of society*. No people have superior courage. They differ from civilized nations only in the manner of exercising it.

Another example of difference of disposition, which proves, in his lordship's opinion, diversity of race, he gives in " the Giagas, a nation of Africa, who bury " all their own children as soon as born, and supply " their places with others stolen from the neighbour- " ing tribes." On this tale I have made the proper comment already. If his lordship's opinion were not well known, we should suspect that he reasoned in this weak manner only to expose to ridicule his favourite doctrine of the difference of species among men. Surely no devotee was ever guilty of more implicit faith than this unbeliever!

The Japanese, his lordship esteems, on this subject, a valuable example. " The Japanese, says he, differ " essentially

* These reasons are well illustrated in Dr. Robertson's history of America.

" eſſentially from the reſt of mankind, becauſe when " others would kill their *enemies*, they kill *themſelves* " through ſpite." If I miſtake not, a native of this ſelf-murdering country might find many of the ſame tribe under London bridge.

The Japaneſe furniſh his lordſhip with another example equally good. " They never ſupplicate the " gods, like other men, in diſtreſs." That difference is certainly very ſtriking, between them and a certain claſs of men who never ſupplicate their Maker at any other time. And yet I have known many Japaneſe, in my time, who have even curſed their Maker, in diſtreſs, as the author of their misfortunes.

His lordſhip acknowledges indeed that theſe arguments are not altogether concluſive; and therefore he proceeds to produce others that he eſteems more perfect in their kind. Theſe I ſhall quote at full length that I may diminiſh nothing of their force; and endeavour to anſwer in as few words as poſſible.

" But not to reſt upon preſumptive evidence, ſays " he, few animals are more affected than men gene-" rally are, not only with change of ſeaſons in the " ſame climate, but with change of weather in the " ſame ſeaſon. Can ſuch a being be fitted for all cli-" mates equally? Impoſſible---horſes and horned cat-" tle ſleep on the bare ground wet or dry without " harm, and yet were not made for every climate: " can a man then be made for every climate, who is " ſo much more delicate, that he cannot ſleep on wet

" ground

" ground withou t the hazard of ſome mortal diſeaſe ?" ---This is the argument. But it is refuted by the whole experience of the world. The human conſtitution is the moſt delicate of all animal ſyſtems : but it is alſo the moſt pliant, and capable of accommodating itſelf to the greateſt variety of ſituations. The lower animals have no defence againſt the evils of a new climate but the force of nature. The arts of human ingenuity furniſh a defence to man againſt the dangers that ſurround *him* in every region. Accordingly we ſee the ſame nation paſs into all the climates of the earth---reſide whole winters at the pole---plant colonies beneath the equator---purſue their commerce and eſtabliſh their factories, in Africa, Aſia, and America. They can equally live under a burning, and a frozen ſky, and inhabit regions where thoſe hardy animals could not exiſt.---It is true, ſuch great changes ought not to be hazarded ſuddenly and without precaution. The greateſt evils that have ariſen from change of climate have been occaſioned by the preſumption of health that refuſes to uſe the neceſſary precautions, or the neglect of ignorance that knows not what precautions to uſe*. But when changes are gradually, and prudently effected, habit ſoon accommodates the conſtitution to a new ſituation, and human ingenuity diſcovers the means of guarding againſt the dangers of every ſeaſon, and of every climate.

But

* Captain Cook has merited great praiſe for the ſervice he has rendered to mankind, by improving the art of preſerving health in long voyages, through the moſt diſtant climates.

But " men, says his lordſhip, cannot ſleep on the " wet ground without hazard of ſome mortal diſ- " eaſe:" and therefore concludes that " they were " not fitted for all climates."---I ſuppoſe by *men* he means Europeans; becauſe the ſavages of America ſleep on the ground without hazard, in every change of weather. Whether, he admits the ſavage into the rank of men or not, he concludes, from this circumſtance, that they are of a different ſpecies from the civilized and poliſhed people of Europe.---If his lordſhip had viſited the foreſts of America he would have found in this, as well as in other inſtances, how little he was acquainted with human nature. He would have ſeen this argument, on which he reſts as a capital proof, totally overturned. He would have ſeen Europeans, or the deſcendents of Europeans, become by habit, as capable as ſavages, of uſing the naked earth for their bed, and of enduring all the changes of an inclement ſky. The Anglo-Americans on the frontiers of the ſtates, who acquire their ſuſtenance principally by hunting, enter with facility into all the habits of ſavages, and endure with equal hardineſs the want of every convenience of poliſhed ſociety*.

So

* Not only the hunters, who have been long uſed to that mode of life, are able to lodge, without injury, on the wet ground, and under all ſeaſons; but the large companies of men, women and children who are continually removing from the interior parts of the United States, to the weſtern countries for the ſake of occupying new lands, encamp, every night, in the open air. They ſleep on the earth, and frequently under heavy ſhowers of ſnow or rain. They kindle a large fire, in the center of their encampment, and ſleep round it, extending their feet towards the pile. And many of them have aſſured me that, while their feet are warm they ſuffer little inconvenience from the vapour of the ground, or even from rain or ſnow.

So that this argument, like all the reſt, is not only inconcluſive to his purpoſe, but militates againſt him.

" But the argument I chiefly rely on, ſays his lordſhip, is that were all men of one ſpecies, there " never could have exiſted, without a miracle, diffe" rent kinds, ſuch as exiſt at preſent. Giving allow" ance for every ſuppoſeable variation of climate, or " of other cauſes, what can follow but endleſs varie" ties among individuals, as among tulips in a gar" den? Inſtead of which we find men of different " *kinds*; the *individuals* of each kind remarkably uni" form, and differing no leſs remarkably from the in" dividuals of every other *kind*. Uniformity without " variation is the offspring of nature, never of chance."

How often do philoſophers, miſtake the eagerneſs and perſuaſion of their own minds for the light of truth and reaſon!---The firſt part of this argument is no more than an ardent and zealous aſſertion. As it reſts on no proof, it needs no refutation. And I confidently appeal to the attentive and reflecting reader to judge, whether I have not aſſigned adequate cauſes of this effect, without the ſuppoſed neceſſity of recurring to miracle.

The ſecond part of this argument, on which ſo much reliance is placed, contains a fine ſimilitude; but that " ſimilitude operates directly againſt his principle. What " can follow, he aſks, but endleſs varieties among in" dividuals, as among tulips in a garden?"---I anſwer, that ſuch varieties among individuals are found in every climate, in every region, in every family. But different climates muſt neceſſarily produce varieties not among *individuals* but among *kinds*. For the ſame climate or the ſame ſtate of ſociety, operating uniformly as far as it extends, muſt produce a certain *uniformity* in the *kind*, and operating *differently* from every other climate, or every other ſtate of ſociety, muſt render that *kind different* from all others.---" Uniformity, ſays he, " is the offspring of nature never of chance." Could his lordſhip mean to inſinuate by this remark that the operations of climate are the effect of chance, or that all its varieties are not governed by uniform and certain laws? Philoſophy is aſhamed of ſuch reaſoning in one of her champions!

He adds, " there is another argument that ap- " pears alſo to have weight, horſes with reſpect to " ſize, ſhape and ſpirit differ widely in different cli- " mates. But let a male and female of whatever cli- " mate, be carried to a country where horſes are in " perfection, their progeny will improve gradually, " and will acquire, in time, the perfection of their " kind. Is not this a proof that all horſes are of " one kind?"

His lordſhip hardly needs an opponent, he reaſons ſo ſtrongly againſt himſelf. The ſpecies of men, no leſs than that of horſes, changes its appearance by every removal to a new climate, and by every alteration of the ſtate of ſociety. The preſent nations of Europe are an example in the way of improvement; the Europeans which he acknowledges have degene-

rated

rated by removing to Africa, Aſia, and South-America, are an example in the contrary progreſſion. Carry the natives of Africa or America to Europe, and mix the breed, as you do that of horſes, and they will acquire in time, the high perfection of the human form which is ſeen in that poliſhed country. Men will acquire it in the ſame number of deſcents as theſe animals. No, ſays his lordſhip, " a mullatto will be " the reſult of the union of a white with a black*."

That is true in the firſt deſcent, but not in the fourth or fifth, in which, by a proper mixture of races, and by the habits of civilized life, the black tinge may be intirely effaced.

There is, at preſent, in the college of New-Jerſey, a ſtriking example of a ſimilar nature, in two young gentlemen of one of the firſt families in the ſtate of Virginia, who are deſcended, in the female line, from the Indian emperor Powhatan. They are in the fourth deſcent from the princeſs Pocahuntis, a high-ſpirited and generous woman. And though all their anceſtors in Virginia have retained ſome characters, more or leſs obvious, of their maternal race, yet, in theſe young gentlemen, they ſeem to be intirely effaced. The hair and complexion, of one of them in particular, is very fair, and the countenance and figure of the face is perfectly Anglo-American. He retains only the dark and vivid eye that has diſtinguiſhed the

* The ſame thing, his lordſhip might have remarked, takes place in horſes as in the human race. The properties of two different breeds, will in the firſt deſcent, be equally blended in th offspring.

the whole family, and rendered ſome of them remarkably beautiful. His lordſhip's argument, therefore, if it be good, is a clear proof againſt himſelf, that all men are of one kind.

He concludes, however, from the preceding remarks which he has made, " that mankind muſt have " been originally created of different ſpecies, and " fitted for the different climates in which they were " placed, whatever change may have happened, in " later times, by war or commerce."

Let us aſk, why *fitted* for the different climates in which they were placed?---The proper anſwer is becauſe they could not exiſt in other climates; or, becauſe they attain the greateſt perfection of their nature only in their own. Both theſe reaſons, in the preſent caſe, are inconſiſtent with experience. Let us remember " the changes that have been produced by " war and by commerce." Nations have tranſplanted themſelves to other climes; yet they continue to exiſt and flouriſh---foreigners have become, aſſimilated to the natives. Inſtead of attaining, in their primitive abodes, the perfection of their nature, they have improved by migrating to new habitations. The Goths, the Moguls, the Africans have become infinitely meliorated by changing thoſe ſkies, for which it is ſaid they were peculiarly fitted by nature. They muſt therefore have defeated, or improved upon the intentions of their Creator; or, at leaſt, have ſhewn the precautions attributed to him, by this author, to have been unneceſſary. Lord Kaims, having endea-

voured

voured to demonſtrate, in the manner we have ſeen, the exiſtence of original varieties among mankind, proceeds to the concluſion in an equal ſtream of cogent reaſoning. " There is a remarkable fact, ſays " his lordſhip, which confirms the foregoing conjec- " tures: as far back as hiſtory goes, the earth was in- " habited by ſavages divided into many ſmall tribes, " each tribe having a language peculiar to itſelf. " Is it not natural then to ſuppoſe that theſe original " tribes, were different races of men placed in proper " climates, and left to form their own language? But " this opinion we are not permitted to adopt, being " taught a different leſſon by revelation. Though we " cannot doubt of the authority of Moſes, yet, his ac- " count of the creation is not a little puzzling. Ac- " cording to that account all men muſt have ſpoken " the ſame language, viz. that of our firſt parents. " But what of all ſeems the moſt contradictory to that " account is the *ſavage ſtate*. Adam, as Moſes in- " forms us, was endued by his Maker, with an emi- " nent degree of knowledge; and he certainly muſt " have been an excellent preceptor to his children, " and their progeny, among whom he lived ſeveral " generations. Whence then the degeneracy of all " men to the ſavage ſtate? To account for that diſ- " mal cataſtrophe mankind muſt have ſuffered ſome " terrible convulſion. That terrible convulſion is re- " vealed to us in the hiſtory of the tower of Babel. " By confounding the language of all men, and ſcat- " tering them abroad upon the face of the earth, they " were rendered ſavages. And to harden them for " their new habitations, it was neceſſary that they

" ſhould

" ſhould be divided into different kinds, ſitted for dif-
" ferent climates. Without an immediate change of
" bodily conſtitution, the builders of Babel could not
" poſſibly have ſubſiſted, in the burning region of
" Guinea, or in the frozen region of Lapland. If
" the common language of men had not been con-
" founded upon their attempting the tower of Babel,
" I affirm that there never could have been but one
" language. Antiquaries conſtantly ſuppoſe a migra-
" ting ſpirit in the original inhabitants of the earth,
" not only without evidence, but contrary to all pro-
" bability. Men never deſert their connexions nor
" their country without neceſſity. Fear of enemies,
" and of wild beaſts as well as the attraction of ſociety,
" are more than ſufficient to reſtrain them from wan-
" dering; not to mention that ſavages are peculiarly
" fond of their natal ſoil."

When ignorance pretends to ſneer at revelation, and at opinions held ſacred by mankind, it is too contemptible to provoke reſentment, or to merit a retaliation in kind.---When a philoſopher deſcends to the diſhoneſt taſk, the moſt proper treatment is to hold out to the world his weakneſs and miſtake. Mankind will heap upon him the contempt he deſerves for intermeddling with a ſubject he does not underſtand. Abſurdity and error are at no time ſo deſpicable as when, in a ridiculous confidence of ſhrewdneſs and ſagacity, they aſſume airs of ſuperiority and ſneer. It would be tedious to remark all the weakneſſes of the paragraph I have juſt quoted. One I will point out, and then I ſhall ſhew, that the whole foundation of

this

this reaſoning is falſe, and indicates an utter ignorance of human nature in that ſtate of ſociety of which he ſpeaks.

" Without an immediate change of bodily conſti-" tution, ſays he, the builders of Babel could not poſ-" ſibly have ſubſiſted in the burning region of Gui-" nea, or the frozen region of Lapland." Yet experience teaches us that mankind can exiſt in every climate. The Europeans, to mention no others, have armies, or colonies, in all the regions of the globe. And if his lordſhip believes that the intenſity of a frozen, or a torrid climate was ſufficient to have deſtroyed the builders of Babel, he ſhould have no objection ſurely, after ſuch a declaration, to acknowledge that they might have altered the figure, or changed the complexion. Yet his whole object is to combat this principle. He allows the greater, he denies the leſs effect. But errors or contradictions of this kind, lord Kaims, in his zeal againſt an obnoxious doctrine, eaſily overlooks.

I propoſed in the next place to ſhew, that the whole foundation on which the reaſoning in this paragraph reſts is falſe, and only proves his ignorance of human nature in that ſtate of ſociety of which he ſpeaks.---It reſts on two principles, 1ſt, That the children of Adam or Noah could never have become ſavage if theſe fathers of the race were the wiſe men which Moſes repreſents them to be---and 2dly, That there never could have exiſted a diverſity of languages. On the other hand, I doubt not of being able to prove that

that the ſavage condition of the greater part of the world was the neceſſary conſequence of one family, and of the ſtate of the earth as Moſes repreſents it immediately after the deluge.--And that out of the ſavage ſtate, diverſity of languages would naturally ariſe.

I am not now going to explain the hiſtory of Babel, or to unfold or defend the miracles recorded in the ſacred ſcriptures. I take the matter on his lordſhip's ground, who, no doubt, moſt devoutly and fervently diſbelieves all miraculous interpoſition of the Deity, and ſhew that, in *the nature of things*, man would become ſavage, and language would become divided.

Man deſcended after the deluge into an immenſe wilderneſs in which the beaſts would naturally multiply infinitely faſter than the human race. Agriculture would probably, from habit and inclination, be the employment of Noah, and his immediate deſcendents; and with them would commence the civilized ſtate which can be traced without interruption, from the countries which they occupied and the period in which they lived, down to our own country, and to the preſent times.---But agriculture furniſhes too ſlow and laborious a ſubſiſtence to be grateful to all men. Many, in the midſt of a wilderneſs filled with beaſts, would be ready to forſake the toils of clearing and cultivating the ground, and to ſeek their proviſion from the chace, which has been ever a favourite exerciſe of mankind, particularly, in rude ages. Hunting would ſoon

ſoon ſpread them over extenſive regions, and diſperſe them widely from one another. Single families, or collections of a few families, ſeated in ſeparate diſtricts of a country almoſt boundleſs, would become independent tribes, and the mode of procuring ſubſiſtence would render them ſavage. His lordſhip ſuppoſes that there is an invincible objection againſt ſuch diſperſion, and ſuch manners, in the example and advice of a venerable anceſtor, and in the ſocial diſpoſition of mankind.---The example and advice of Noah and his ſons would doubtleſs have great influence on that civilized people, which would naturally grow up round their immediate habitation. But how ſhould they influence their remote deſcendents who were ranging the foreſts at the diſtance of an hundred or a thouſand leagues? To anſwer this queſtion, he confidently pronounces that mankind would always have been within the reach of this example, becauſe they never would have ſeparated from one another, and from the pleaſures of improved ſociety.---" Men, ſays he, " never deſert their connexions, nor their country " without neceſſity.---fear of enemies, and of wild " beaſts, as well as the attractions of ſociety, are more " than ſufficient to reſtrain them from wandering: " not to mention that ſavages are peculiarly fond of " their natal ſoil."

Theſe ideas are derived from civilized ſociety, and are not applicable to ſavage life. 'Tis ridiculous to talk of the fear of wild beaſts to men whoſe diverſion it is to purſue and ſlay them---and not much leſs abſurd is it, to talk of the attractions of ſociety, and of attachments to a natal ſoil, to people in a wilderneſs, to whom migration is a habit---to whom every ſpot of ground is equal where they can find game---and who feel the charms of the chace more ſenſibly than the charms of ſociety. What is the pleaſure of ſociety in that rude ſtate?---Deſtitue of ſentiment, or converſation, it is little more than the pleaſure that dumb animals feel at the approach of other animals of the ſame ſpecies. The chace, which to them is productive of higher and ſtronger enjoyments, eaſily breaks the feeble ties of ſuch ſociety; and hunters, like beaſts of prey, delight in ſolitudes and deſerts.--- Men in ſuch a ſtate migrate through caprice, or through curioſity, or for the convenience of hunting.---The influence of extenſive lands lying in common, and ready to be occupied by the firſt comer, is extremely viſible on the inhabitants of theſe United States. Their fathers came from Europe with all thoſe fixed habits, and thoſe tendencies to local attachments which can reaſonably be imputed to any people. They took poſſeſſion of a boundleſs foreſt, which had a ſpeedy and an aſtoniſhing effect on their manners. The Anglo-Americans diſcover comparatively little attachment to a native ſoil. No hereditary poſſeſſion, no objects of antiquity ſeize the imagination, and fix it to a certain ſpot. The people, migrate without reluctance, to the greateſt diſtances---they change their habitations as ſoon as they become ſtraitened in their quarters---and forſake their friends, and the place of their nativity, for apparently ſmall conveniencies. This is more the caſe as you paſs from the cultivated lands near the ocean, towards the weſtern frontiers.

In

In proportion as the citizens of the ſtates approach the vicinity of the Indian tribes, ſimilarity of ſituation produces alſo a great approximation of manners. If his lordſhip had ſeen America, he would have ſeen men forever migrating from the midſt of ſociety to uncultivated deſerts---he would have often ſeen them forſake the conſtraints of civilization, for the independence, and the charms of a ſtate approaching to ſavage---he would have ſeen the frontiers of all the United States filled with the deſcendents of Europeans, who have, in a great meaſure, adopted the manners of the native Indians, along with their mode of procuring ſubſiſtence---he would have ſeen theſe people, as ſociety advances upon them from the cities, and the ſea-coaſt, retreating before it into the wilderneſs---he would have ſeen men decline the labours of agriculture as a toil, and prefer the fatigues of hunting to all other pleaſures---he would have ſeen that mankind often find charms in the indolence and independence of the ſavage ſtate ſuperior to thoſe that reſult from the refinements and attractions of civil ſociety, which muſt be purchaſed with labour, and held by ſubordination---he would have ſeen that wanderers have no *attachment*, as he ſuppoſes, *to their natal ſoil*---he would have ſeen multitudes of the people of theſe United States, change their habitations without regret---he would have ſeen the Indians, either ſingly, or in companies, travel for many moons ſucceſſively, to explore other foreſts, and to ſeek for other rivers---he might have ſeen whole tribes riſe from their ſeats at once, and carrying with them the bones of their fathers, ſeek new habitations at the diſtance of an hundred

dred or two hundred leagues.---But his lordſhip has not ſeen them, and he ſpeaks of the ſavage ſtate without underſtanding it, and of human nature, in the beginning of time, without knowing how it would operate then, or how it has operated, in ſimilar ſituations, in later periods. Like many other philoſophers he judges and reaſons, only from what he has ſeen in a ſtate of ſociety highly improved; and is led to form wrong concluſions from his own habits and prepoſſeſſions. On his principles, a ſavage ſtate could never have exiſted, on the ſuppoſition of many races of men, more than of one. *Fear of wild beaſts*, and *the attractions of ſociety* would have held each race together and *prevented their diſperſion*. Everv art of agriculture would have been tried, before they would have extended their habitations into the *dangerous* wilderneſs. A civilized community would have ariſen round their firſt habitations. And when they ſhould have been compelled by neceſſity to enlarge their limits, they would have done it in ſociety. The foreſt would have fallen before them as they advanced; and fear and the ſocial principle would have equally contributed to reſtrain them from the hazards, and the diſperſion conſequent upon the ſpirit of the chace. The world, inſtead of being filled with numerous tribes of ſavages, would have every where preſented to us civilized and poliſhed nations. His lordſhip, on this ſubject, forever reaſons againſt himſelf. He means to combat the doctrine of one race by the exiſtence of the ſavage ſtate; which yet, is a neceſſary conſequence of that doctrine, and would be certainly precluded on his own principles.

His

His lordſhip's next error conſiſts in aſſerting that, " on the ſuppoſition of one race, there never could have exiſted a diverſity of languages". This error is the conſequence of the preceding. Both principles are intimately connected together. Similarity of language would naturally have ariſen out of univerſal civilization, continued down from the original of the race. Diverſity of language neceſſarily ſprings out of the ſavage ſtate. The ſavage ſtate has few wants, and furniſhes few ideas that require *terms* to expreſs them. The habits of ſolitude and ſilence incline a ſavage rarely to ſpeak. When he ſpeaks it is chiefly in figures, and the ſame terms are uſed for different ideas*. Speech muſt, therefore, be extremely narrow, in this rude condition of men. It muſt, likewiſe, be extremely various. Every new region, and every new climate will preſent different ideas, and create different wants, that will naturally be expreſſed by various terms. Hence will originate great diverſity in the firſt elements of ſpeech among all ſavage nations.

Savages ſpeak ſo much by figure, and even by geſture, that it greatly contracts the limits of their language. They have no adjectives, no particles, no abſtract terms, no ſingular denominations. They have no parts of ſpeech but the ſubſtantive and the verb. Their verbs are confined to a very few ſtates and actions of animals; and perhaps ſome other objects of nature that are moſt familiar. Their ſubſtantives conſiſt of a few general names of animals, of vegetables and of ſome of the moſt obvious parts of the inanimate world, ſuch as rocks, rivers, mountains. When they would expreſs a quality, they do it figuratively by applying the name of one ſenſible object to another. A deer is a ſwiſt man---a fox is a wiſe or an artful man ---a bear is a ſtrong, a furious, or a courageous man. Thus by applying the ſame term to ſignify ſeveral ideas, by having but two parts of ſpeech, and theſe derived from few objects, and by uſing geſtures frequently to ſupply the place of the verb, ſpeech is reduced, among them, to a narrow compaſs.

nations. If a few common principles ſhould be handed down from the original family; yet theſe, in time, would be changed by the uſual flux of language. Tongues would become as various as the tribes of men. Speech being, therefore, in the firſt ages, both extremely narrow, and extremely diverſified, theſe rude people would begin their progreſs towards improvement, with few, or with no elements in common. And in the infinite multitude of words which civilization and refinement add to language, no two nations, perhaps, have ever agreed upon the ſame ſounds to repreſent the ſame ideas. Superior refinement, indeed, may induce imitation, conqueſts may impoſe a language, and extenſion of Empires may melt down different nations, and different dialects into one maſs. But independent tribes naturally give riſe to diverſity of tongues. Thus, perhaps, the ſpeech of men was at firſt one---it became gradually divided into a multitude of tongues---and the progreſs of civilization, and the mixing of nations by conqueſt or by commerce, tends to bring it back again towards one ſtandard.--- His lordſhip fails in every proof. And this laſt argument, which he deemed among the ſtrongeſt, againſt the hiſtory of the ſcriptures, and the common origin of mankind, militates like the reſt againſt himſelf, and confirms the doctrine that he oppoſes.

Such is the attack which this celebrated philoſopher has made on the doctrine of one race. In all the writings of this author, there is not another example of ſo much weak and inconcluſive reaſoning. This ought in juſtice to be imputed to the cauſe, and

not

not to the writer. His talents are univerſally acknowledged. It was for that reaſon I choſe to make theſe ſtrictures on him, rather than on an author of inferior name. He has probably ſhewn the utmoſt force of that cauſe which he has undertaken to defend. If he has failed, it is only becauſe it is incapable of defence. For, to him I may apply the lines which, on another ſubject, he applies to Dr. Robertſon.

——————————————*Si Pergama dextrâ*
Defendi poſſent, etiam hâc defenſa fuiſſent.

THE END.

www.ingramcontent.com/pod-product-compliance
Lightning Source LLC
Chambersburg PA
CBHW030601270326
41927CB00007B/997
* 9 7 8 3 3 3 7 7 8 0 8 0 7 *